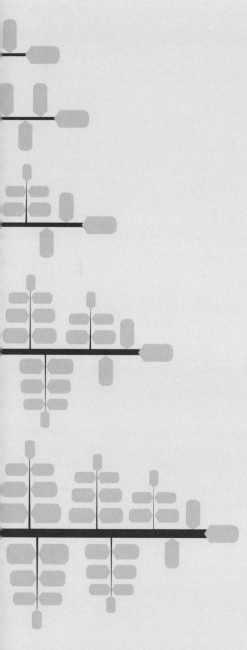

ferns
and lycophytes
OF AOTEAROA NEW ZEALAND

Deparia petersenii subsp. *congrua*, page 136

IDENTIFICATION GUIDE TO THE
ferns
and lycophytes
OF AOTEAROA NEW ZEALAND

Leon Perrie & Patrick Brownsey

TE PAPA

PRESS

IN MEMORY OF PATRICK BROWNSEY, 1948–2023

Cyathea medullaris, page 130.

Asplenium bulbiferum, page 77.

Contents

Lecanopteris pustulata, page 187.

Introduction

Ferns and lycophytes in Aotearoa New Zealand

Aotearoa New Zealand is internationally acclaimed for its unique flora, with the prominence of ferns a feature of many of our quintessential landscapes. Such is the conspicuous presence of ferns that 'Fernland' was an early colloquial name by British settlers for New Zealand. Today, ferns are cultural and commercial icons of Aotearoa, particularly the ponga silver fern (page 129), which is the country's national nature emblem alongside the kiwi.

In wetter forests, tall tree ferns fill the sub-canopy and filmy ferns clothe the ground, trunks and branches. Other species perch or climb on trees, and ground ferns can be abundant and diverse. When forest is disturbed or even cleared, be it by slips or fire or machinery, ferns such as the mamaku tree fern (page 130) and rārahu bracken (page 238) can play a prominent role in its recovery. Other species of ferns and lycophytes are specialised for wetlands, the coast or alpine areas. There are even ferns that prefer drier, open environments, and some indigenous species flourish within urban areas.

With 204 species, ferns and lycophytes comprise about 8 percent of the indigenous vascular flora of Aotearoa. That number of species is low compared to the typical biological richness of tropical regions. The likes of Fiji and Solomon Islands are home to many more species than Aotearoa despite their much smaller landmasses. However, about 44 percent of the indigenous ferns and lycophytes in Aotearoa are endemic, found only here. This high endemism stands Aotearoa apart from its Pacific island neighbours (although New Caledonia, another part of the otherwise largely submerged continent of Zealandia, closely follows). The remaining 56 percent are also indigenous to other lands, particularly south-eastern Australia, reflecting proximity and the somewhat similar temperate climate. Many species are indigenous to both Aotearoa and Tasmania, Victoria, or southern New South Wales, with our ferns embodying a strong trans-Tasman relationship.

Although ferns and lycophytes have an ancient fossil record, it seems that most of the living species in Aotearoa are derived from ancestors that arrived only after the landmass had geologically separated from the rest of Gondwana. This is supported by genetic studies of relationships combined with the global fossil record, and it explains why there are so many species shared by Aotearoa and elsewhere – their populations look the same because they share a relationship that is much more recent than the several tens of millions of years of geological isolation of Aotearoa. The microscopic, wind-blown spores of ferns and lycophytes have meant that the seas surrounding Aotearoa have not been a complete barrier to their migration. Given the respective sizes of the landmasses and the prevailing westerly winds, much of the movement probably has been immigration to Aotearoa, from Australia but also elsewhere. However, there is genetic evidence that Aotearoa has also been a source of emigration for at least a few Australian populations of species shared with New Zealand, and perhaps for the rest of the world in the case of the *Hymenophyllum* filmy ferns.

Many exotic ferns have been introduced to Aotearoa by human activities. Some of these have become problematic weeds, displacing indigenous species, curtailing economic productivity and even transforming environments. While 'delicate' may be a popular description for ferns, many species, both indigenous and exotic, are tough competitors.

Learning how to distinguish different fern and lycophyte species is an initial step in connecting with this rich element of our flora and unlocking the mātauranga and other knowledge that surrounds them. With ferns and lycophytes all around us, this book can facilitate that connection wherever you are.

What are ferns and lycophytes?

Ferns and lycophytes are two of the six principal groups of extant land plants, the others being seed plants, liverworts, hornworts and mosses. Ferns are most closely related to seed plants (including flowering plants and conifers), although it is a distant relationship as they diverged several hundred million years ago. The next closest relatives are the lycophytes.

These three groups – ferns, seed plants and lycophytes – are collectively known as vascular plants because of their 'plumbing', provided by xylem and phloem cells, which conveys water and nutrients around their bodies. The liverworts, hornworts and mosses – collectively known as bryophytes – generally lack 'plumbing', and their interrelationships remain unclear. Microscopic, single-celled and usually wind-borne spores are the principal means of dispersal for ferns, lycophytes and bryophytes, whereas seed plants have multi-cellular seeds.

Ferns and lycophytes are similar in being vascular plants that disperse by spores. However, they differ in that ferns, like seed plants, have big leaves with branching veins, whereas lycophytes, like many mosses, have generally smaller leaves with a single unbranched vein. Ferns usually have frond-like leaves (exceptions include *Equisetum* and the Psilotaceae) that are tightly coiled when young; these tight coils are

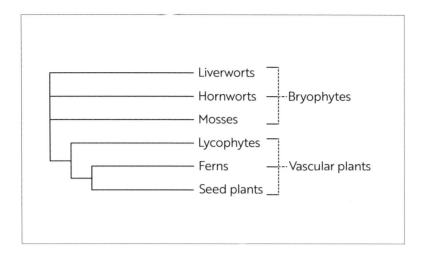

known as koru in Aotearoa and croziers or fiddleheads elsewhere. Some seed plants also have fern-like leaves, but any plant with flowers or fruits is not a fern. Similarly, some bryophytes might be confused with small ferns, but their distinctive, usually stalked, spore-producing capsules are quite different from the reproductive structures of ferns.

ABOUT THIS GUIDE

This book is designed as an introduction to the identification of ferns and lycophytes in Aotearoa New Zealand, suitable for anyone interested in the diverse array of ferns and lycophytes that make the landscape of Aotearoa so unique. Designed to be both taken out into the field and consulted for further information, it aims to give readers a solid foundation for identification. It summarises the more technical and comprehensive accounts given in *Flora of New Zealand – Ferns and Lycophytes*, hereafter 'the eFloraNZ', available at www.nzflora.info.[1]

    This guide features 201 species of ferns and lycophytes (three with two subspecies each) most likely to be encountered in the wild land habitats of Aotearoa. These belong to 64 genera and 29 families. There are 15 featured species of lycophyte, in the families Isoetaceae, Lycopodiaceae and Selaginellaceae (their genera are listed in Appendix 1). The remaining featured

---

1. The eFloraNZ, compiled by the authors, provides accounts for all 204 indigenous and 64 naturalised species recorded from the New Zealand Botanical Region; it also includes identification keys for all families, genera and species as well as a detailed introduction to the ferns and lycophytes of Aotearoa New Zealand. See Brownsey, PJ and LR Perrie, Introduction, in I Breitwieser (ed.), *Flora of New Zealand – Ferns and Lycophytes*, Fascicle 1, Manaaki Whenua Press, Lincoln, 2022.

species are all ferns. Most species are indigenous to Aotearoa; however, 22 of the exotic species that have become naturalised are also included.

Species found only on Rangitāhua Kermadec Islands, Manawatāwhi Three Kings Islands, Rēkohu Chatham Islands or subantarctic islands are not included, nor are several very localised or otherwise uncommon species. The 24 indigenous species not included in the **Species profiles** are listed in Appendix 2.

This book has four main components:

1. **Illustrated glossary** (pages 19–29). A guide to the terms used to describe the various structures and characteristics of the plants. Readers should familiarise themselves with these terms and refer back to them as needed in order to successfully identify a species.

2. **Guide to genera** (pages 31–41). A picture-based guide to aid in genus identification. Each photo shows the reproductive structures that characterise the genus and is accompanied by brief descriptions of additional distinguishing features.

3. **Species scan** (pages 43–61). A reference photo of each species to enable a quick scan for likely candidates. The images are arranged by evolutionary relationships between genera and families, meaning that similar looking species are usually on the same or neighbouring pages.

4. **Species profiles** (pages 63–267). A detailed account for each species, arranged alphabetically first by genus and then by species. Each profile includes:

   - name(s); naturalised species (denoted [†]) are indicated following their accepted scientific name

   - family within the taxonomic classification

   - photos showing overall appearance and diagnostic characteristics

   - a distribution map for Te Ika-a-Māui North Island, Te Waipounamu South Island and Rakiura Stewart Island[2]

   - global distribution, including whether the species is *endemic* (found only in Aotearoa New Zealand), *indigenous* to Aotearoa and other areas, or *naturalised* (an exotic species introduced via human activities and now self-reproducing in Aotearoa)

---

2. Distribution data are based on herbarium specimens verified for the eFloraNZ, with occasional additions.

Species profile page features

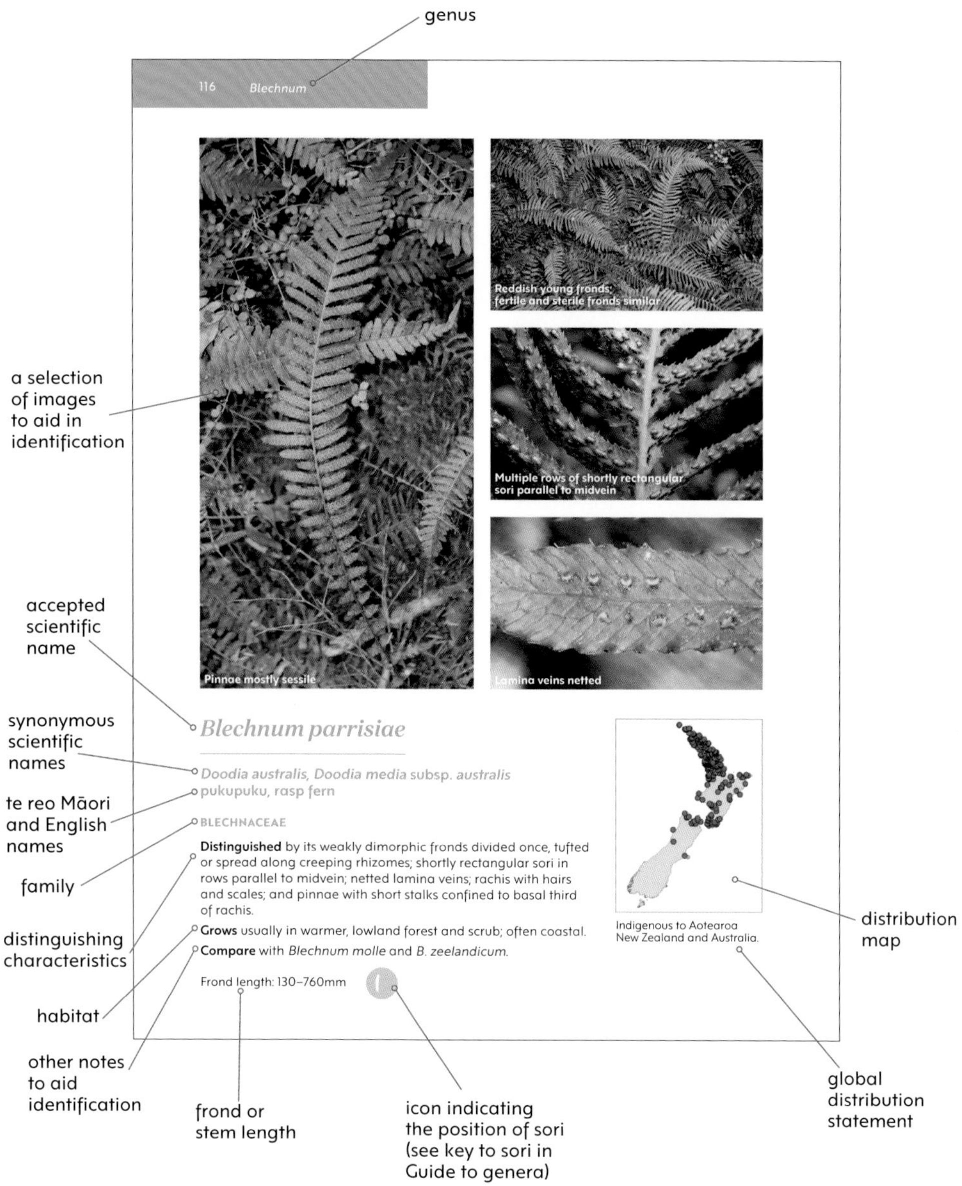
genus
116 Blechnum
Reddish young fronds; fertile and sterile fronds similar
Multiple rows of shortly rectangular sori parallel to midvein
Pinnae mostly sessile
Lamina veins netted
a selection of images to aid in identification
accepted scientific name
synonymous scientific names
te reo Māori and English names
family
distinguishing characteristics
habitat
other notes to aid identification
Blechnum parrisiae
Doodia australis, Doodia media subsp. australis
pukupuku, rasp fern
BLECHNACEAE
Distinguished by its weakly dimorphic fronds divided once, tufted or spread along creeping rhizomes; shortly rectangular sori in rows parallel to midvein; netted lamina veins; rachis with hairs and scales; and pinnae with short stalks confined to basal third of rachis.
Grows usually in warmer, lowland forest and scrub; often coastal.
Compare with Blechnum molle and B. zeelandicum.
Frond length: 130–760mm
Indigenous to Aotearoa New Zealand and Australia.
distribution map
global distribution statement
frond or stem length
icon indicating the position of sori (see key to sori in Guide to genera)

- distinguishing characteristics that are recognisable in the field

- a brief account of the species' habitat

- other notes, including similar species in different genera, likely confusable species within large genera and previously misapplied names

- the range in frond or stem length of fertile mature plants[3]

- an icon indicating the position of the sori (as used in the **Guide to genera** section).

Browsing the photos in the **Species scan** and **Species profiles** sections will facilitate identification in many instances. Another approach is to use the **Guide to genera** to pick out candidate genera for further inspection. It may be necessary to browse several genera; similar-looking species will only be successfully identified by checking the distinguishing characteristics and maps on their respective pages.

Regardless of the approach, it is best to review all species within a genus and its family (see Appendix 1). Species from different genera that are easily confused are given within the other notes section of the **Species profiles**.

Reproductive structures (such as the presence of sori) found on spore-producing fertile fronds are important when using the **Guide to genera**, and often for species identification in general. If the plant found is not fertile, it is worth searching for others that are, as identification may otherwise be impossible. Generally, bigger individuals are most likely to be reproducing; small individuals may simply be immature. The reproductive structures of most ferns are situated on the underside of their fronds, while for lycophytes they are usually in cones. Tree ferns belong to the genera *Cyathea* or *Dicksonia*. When they are over about 2m tall, even if their reproductive structures are out of reach, they can usually be identified by other characteristics of their fronds and trunks.

Other characteristics that should be noted are:

- the division of the whole frond. Working from only a fragment can be misleading;

---

3. Ranges given are as recorded in the eFloraNZ and do not encompass juveniles or two species that scramble extensively. Measurements from Orchard are used for two species naturalised from Australia that were not measured for the eFloraNZ. See Orchard AE, *Flora of Australia Volume 48 – Ferns, gymnosperms and allied groups*, ABRS/CSIRO, Melbourne, 1998.

- whether the fronds are hairy or scaly. Hairs and scales are best seen on young fronds; they fall off as fronds age;

- whether the fronds are tufted or spread along creeping rhizomes.

A ×10 hand lens or magnifying glass is recommended, particularly when examining reproductive structures or hairs and scales.

The **Steps to identifying a fern or lycophyte** flowchart on the following page can be consulted as a quick reference on how to make use of the different sections of the book when identifying a species.

Remember that plants in reserves are protected and should not be removed – in part or whole – without permission. Instead, take photos of the reproductive structures and other characters listed above.

If you believe you have found a fern outside of its mapped distribution, please upload photos to www.iNaturalist.nz and email leon.perrie@tepapa.govt.nz.

Names used

The accepted scientific name for each species is used in the heading on the relevant species page and follows the eFloraNZ.[4] It comprises a genus name (first) and a species name (second). Subspecies (abbreviated subsp.) are named except where only one of multiple subspecies occurs in Aotearoa New Zealand. Closely related genera are aggregated into families. Appendix 1, lists the genera in each family.

Synonymous scientific names that have been used recently, particularly within New Zealand resources, are given directly below the main heading. These are alternative names in different genera, or redundant names.

Te reo Māori names follow Beever[5], supplemented by Te Aka (www.maoridictionary.co.nz). English names are from the eFloraNZ, except for a handful here coined. While many species have no colloquial name, some have many, in either or both languages. Generally, only one name in te reo Māori and English is given, with the selection following Beever for te reo and author judgement for English.

Te reo Māori and English names applied collectively to a genus or family are given in Appendix 4.

---

4. The approach to taxonomic change taken by the eFloraNZ was more conservative than for many other authorities, with many arguably superfluous changes to generic placement not accepted.
5. Beever, J, *A dictionary of Maori plant names*, Auckland Botanical Society, Auckland, 1991.

STEPS TO IDENTIFYING A FERN OR LYCOPHYTE

Finding reproductive structures on fertile fronds or stems is key in fern and lycophyte identification. Generally, larger plants are more likely to be fertile. Identification is usually best done while looking at a whole plant. A magnifying glass or hand lens is recommended to aid in successfully identifying reproductive structures.

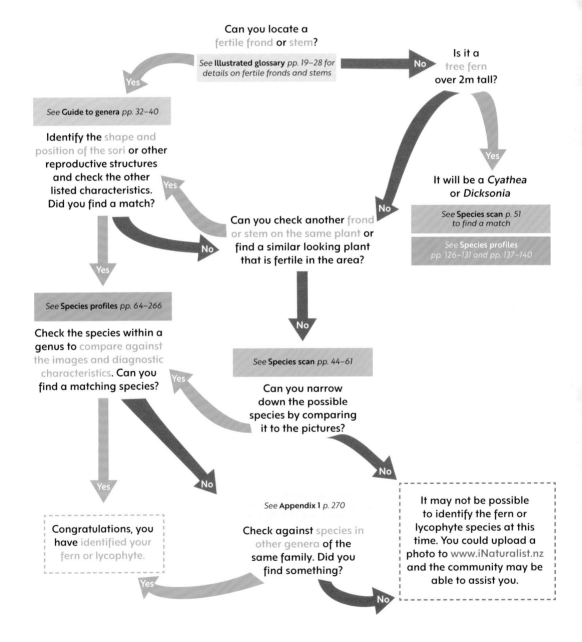

Can you locate a fertile frond or stem?

See **Illustrated glossary** *pp. 19–28 for details on fertile fronds and stems*

No

Yes

Is it a tree fern over 2m tall?

See **Guide to genera** *pp. 32–40*

Identify the shape and position of the sori or other reproductive structures and check the other listed characteristics. Did you find a match?

Yes

No

Yes

It will be a *Cyathea* or *Dicksonia*

See **Species scan** *p. 51 to find a match*

See **Species profiles** *pp. 126–131 and pp. 137–140*

Yes

Can you check another frond or stem on the same plant or find a similar looking plant that is fertile in the area?

No

No

See **Species profiles** *pp. 64–266*

Check the species within a genus to compare against the images and diagnostic characteristics. Can you find a matching species?

No

Yes

See **Species scan** *pp. 44–61*

Can you narrow down the possible species by comparing it to the pictures?

No

Yes

Yes

No

No

See **Appendix 1** *p. 270*

Congratulations, you have identified your fern or lycophyte.

Check against species in other genera of the same family. Did you find something?

Yes

No

It may not be possible to identify the fern or lycophyte species at this time. You could upload a photo to www.iNaturalist.nz and the community may be able to assist you.

Dicksonia fibrosa, page 137.

Deparia petersenii subsp. *congrua*, page 136.

Illustrated glossary

This glossary defines important terms used in identification. In the **Species profiles**, key characteristics of each species of fern and lycophyte are generally described in plain language. This inevitably involves some loss of precision, but the eFloraNZ can be consulted where more detail is required.

The headings below refer to parts/characters of ferns and lycophytes, whereas terms in **bold** refer to character states.

Rhizome

The stem of a fern. Often underground.
The rhizome may be **erect**, producing fronds in a **tuft**. Erect rhizomes can be elongated into a trunk as with tree ferns, but they usually do not protrude far above the ground or other surface on which they are growing.

Alternatively, the rhizome may be **creeping**, producing fronds spread along it. Some creeping rhizomes are climbing.

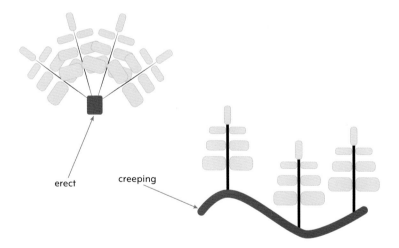

erect   creeping

Frond

The leaf of a fern, comprising its axes (stipe, rachis, costae and midveins) and lamina (shown in the diagram on page 20 and discussed in more detail below).

Fronds can range from **undivided** (entire) to **divided** four or more times. Technical terms exist to describe precisely the degree of

Frond and its parts

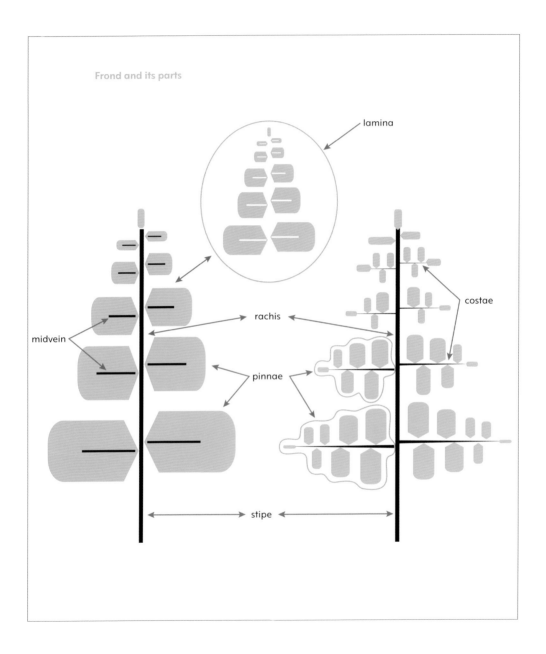

division (see the eFloraNZ). They are not used here, which means some leeway needs to be allowed when using this character.

The degree of division given in the guide is generally the maximum shown by a frond. 'Divided 2–3 times' means that the maximum division of a species varies between divided twice and divided thrice. 'Fronds divided twice (barely three times)' denotes that a species has mostly twice divided fronds but some parts will be divided thrice if closely inspected.

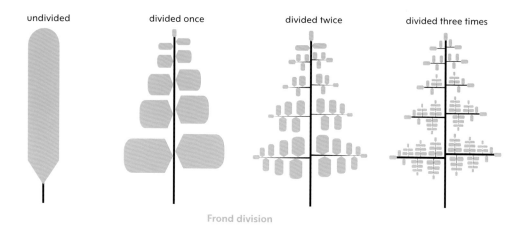

undivided   divided once   divided twice   divided three times

Frond division

Primary, secondary, tertiary, etc. are terms describing the degree of division. See entries for costa and pinna.

Fronds can have **dimorphic** fronds, when different-looking **sterile** and **fertile** (spore-bearing) fronds are produced. This occurs in most *Blechnum* species.

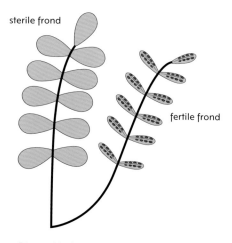

sterile frond

fertile frond

Dimorphic fronds

Rachis

The main, central axis of the frond within the **lamina**.

Stipe

The main axis of the frond below the branching of the primary pinnae, or for undivided fronds below the **lamina**.

Axes like the stipe and rachis can be **winged**, with narrow strips of green tissue along their sides, for part or all of their length.

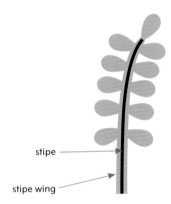

stipe

stipe wing

Winged stipe

Costa (costae)

Primary costae are the subsidiary axes branching from the rachis. These may in turn branch to give secondary costae, etc.

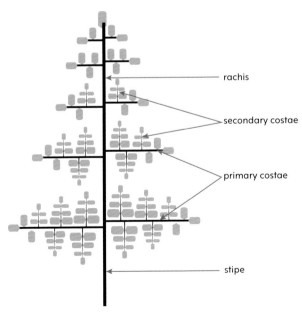

rachis

secondary costae

primary costae

stipe

Lamina (laminae)

The 'leafy' part of the frond, attached to and supported by the axes.

Pinna (pinnae)

A subpart of the lamina that is divided to the rachis (in primary pinnae) or to the relevant costa (in secondary or tertiary pinnae). The smallest subparts can be called lamina segments.

A primary pinna comprises a primary costa (or midvein) and its lamina(e), and may in turn be divided into secondary pinnae, etc. Primary pinnae attach to the rachis, secondary pinnae attach to a primary costa, tertiary pinnae attach to a secondary costa, etc.

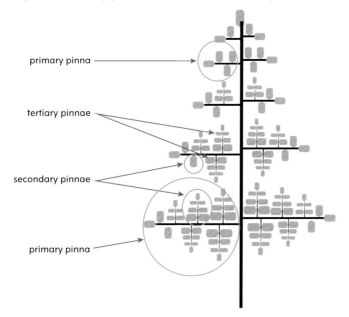

Pinnae and segments may be **stalked** or **sessile** (without stalks). Pinnae can be opposite with pairs of pinnae originating from the same place on each side of the rachis, or alternate where the pinnae are not in opposite pairs.

Lobed means that the relevant **lamina** segment is divided, but not all the way to the **rachis** or **costa**.

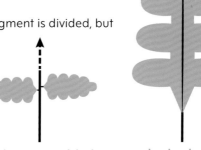

Lamina segments lobed
(lobed primary pinnae)

Lamina deeply
once-lobed

Midvein

The central vein within an undivided frond or lamina segment.

The smaller veins branching from the midvein can be **free**, where they may fork but do not reunite between the midvein and the segment margin. Or they may be **netted**, with the smaller veins forking and reuniting.

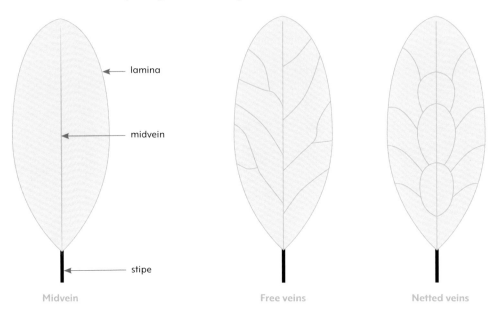

lamina

midvein

stipe

Midvein   Free veins   Netted veins

Margin

The margin of the **lamina** can be **toothed** or **entire** (not toothed). The margin may also be **hairy**, and sometimes bears the reproductive structures.

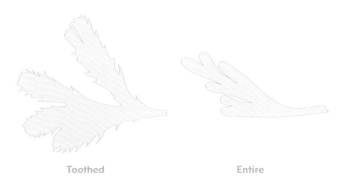

Toothed   Entire

Upperside, underside

Generally used to describe the surfaces of a more-or-less horizontal frond with respect to the ground. Used in this guide in place of the terms adaxial (upperside) and abaxial (underside) used in the eFloraNZ.

Apex (or apical)

The part of a **lamina**, **costa**, etc., furthest from its point of origin.

Base (or basal)

The part of a **lamina**, **costa**, etc., closest to its point of origin.

Hairs and scales

Hairs are narrow (one cell wide), with an appearance like mammalian hairs.

Scales are wider (multiple cells wide) and are usually clearly broader than mammalian hairs.

Where there are no hairs or scales, this is described as **glabrous**.

Hairs can be **glandular**, with a spherical apex, or **non-glandular**, often with a pointed **apex**.

Hairs and scales can be **spreading** from their surface of attachment, or **appressed** where they are closely pressed against the surface. Fronds and leaves can be similarly arranged.

Glandular hair   Non-glandular hair   Scale

Reproductive structures

sporangia

indusium

sorus

Sporangia (sporangium)

The usually spherical capsules that contain the spores by which ferns reproduce and disperse.

Sori (sorus)

Sporangia are often aggregated into discrete clusters known as sori, found on the lamina underside or at its margin. The position of the sori (e.g., away from, or at, the lamina margin) and their shape (e.g., linear or circular or tubular) help to distinguish fern families and genera.

The **colour** of sporangia, and their sori, can change with development. They are often whitish when immature, black when the spores are ripe, and brown/orange after the spores are shed. Additionally, indusia (see below) may crumple with age.

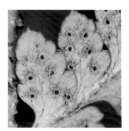

immature spores   mature spores   dispersed spores

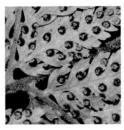

sori elongated and away from lamina margin

sori elongated on lamina margin

sori circular, away from lamina margin

sori circular, on lamina margin

Indusia (indusium)

Structures found in some ferns that partially cover and protect the sori. Their shape can help distinguish genera. Pseudo-indusia formed from an inrolled lamina margin are also included here.

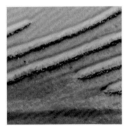

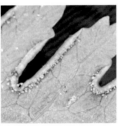

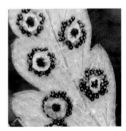

Indusia in the form of an elongated flap opening towards the midvein.

Indusia of the lamina margin, inrolled for short lengths.

Indusia of the lamina margin, inrolled for long lengths.

Circular indusia held on a stalk above circular sori.

Kidney-shaped indusia held on a stalk above circular sori.

Twin indusia, one being an inrolled lamina margin and the other arising from the interior, together enveloping each spherical sorus sited at the lamina margin.

Indusia in the form of a tube.

Sporophylls

Used to refer to the fertile leaves of lycophytes and the fern families Psilotaceae and Equisetaceae. Sporophylls in these groups usually bear sporangia on their upper surface and are usually small with a single, unbranched vein.

Sporophylls with brown sporangia

Sporophylls with white sporangia

Cones

Sporophylls sometimes look different from sterile leaves and are often grouped into cones of varying distinctiveness. Cones may be clearly **stalked** or **sessile**.

Stalked cone

Sessile cone

Bulbils

Vegetative offshoots capable of forming new plants.

Bulbils emerging on frond   Bulbils emerging on stem

Stolons

Creeping horizontal stems that can give rise to new plants.

Blechnum fluviatile, page 107.

Dicksonia squarrosa, page 140.

Guide to genera

Guide to genera

This section will help readers identify the genus to which each fern belongs.

Using the shape and position of the sori on a fertile plant, determine which of the following groups the plant belongs to. Then scan the images and read the characteristics for the genera within that group on the following pages.

 Sori elongated on lamina margin

 Sori short or circular or tubular on lamina margin

Note that some genera (denoted *) occur in more than one of these groups. Also, a few genera are not distinguishable using this approach and are grouped together in this section but their species are presented separately in the **Species profiles** section.

The genera in each group are in alphabetical order except for some instances where similar-looking genera have been brought together to aid comparison.

 Sori elongated and away from lamina margin

 Sori short or circular, and away from lamina margin

 Sporangia not in discrete sori of regular shape and size, lacking indusia

 Stems with single-veined leaves

 Other

SORI ELONGATED ON LAMINA MARGIN

Adiantum *

» Fronds divided 1–5 times, with hairs or glabrous, at least somewhat spread along creeping rhizomes.
» Lamina segments fan-shaped or oblong.
» Indusia comprising short sections of inrolled lamina margin.

Asplenium *

» Fronds divided 2–4 times, usually tufted.
» Laminae with scales, often sparse.
» Sori diagonal to one another and on margin closest to costa.

Blechnum *

» Fronds divided once, rarely twice; tufted or spread along creeping rhizomes.
» Fertile fronds dimorphic, usually darker with much narrower pinnae.

Cheilanthes

» Fronds usually small and divided 2–3 times; on creeping rhizomes but can appear tufted.
» Laminae with scales and hairs, but sometimes barely so.
» Sori in short sections or continuous.

Histiopteris

» Fronds often large, divided 2–4 times, spread along creeping rhizomes.
» Laminae glabrous, with netted veins.

Lindsaea

» Fronds usually small, divided 1–4 times, tufted or spread along creeping rhizomes.
» Indusia opening outwards.

Paesia

» Fronds divided 3–4 times, spread along creeping rhizomes.
» Fertile lamina segments somewhat triangular.
» Laminae with hairs, sometimes sticky, with free veins.

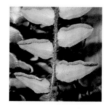

Pellaea

» Fronds divided once, spread along creeping rhizomes.
» Rachises with scales.

Pteridium

» Fronds often large, divided 2–5 times, spread along creeping rhizomes.
» Laminae tough and inconspicuously hairy, with free veins.

Pteris

» Fronds divided 1–5 times, tufted in most species.
» Laminae appearing glabrous, with netted or free veins.

SORI SHORT OR CIRCULAR OR TUBULAR, ON **LAMINA MARGIN**

Adiantum *

» Fronds divided 1–5 times, with hairs or glabrous, tufted or spread along creeping rhizomes.
» Lamina segments fan-shaped or oblong.
» Indusia kidney-shaped, formed from inrolled lamina margin.

Dicksonia

» Fronds large, divided 3–5 times, with hairs, tufted and usually on a trunk.
» Indusia of inner and outer flaps enveloping spherical sori.

Hiya

» Fronds divided 2–3 times, with hairs, spread along creeping rhizomes.
» Lamina veins reaching margin in small indentations.
» Indusia formed from inrolled lamina margin.

Hypolepis *

» Fronds divided 3–5 times, with hairs, sometimes glandular, spread along creeping rhizomes.
» Indusia absent or formed from inrolled lamina margin.

Hymenophyllum

» Fronds usually small, undivided to divided five times, spread along creeping rhizomes (tufted in one species).
» Laminae usually translucent, occasionally covered with hairs.
» Sori usually within indusia of two flaps.

Hymenophyllum

» Fronds undivided, round or kidney-shaped, spread along creeping rhizomes.
» Laminae translucent.
» Sori in cup-like indusia.

Loxsoma

» Fronds divided 3–4 times, spread along creeping rhizomes.
» Lamina not translucent, underside whitish or green.
» Sori in tube-like indusia.

Trichomanes

» Fronds small, divided 1–4 times, tufted or spread along creeping rhizomes.
» Laminae translucent.
» Sori at lamina margins in tubular or trumpet-like indusia.

SORI ELONGATED AND AWAY FROM LAMINA MARGIN

Anogramma
» Fronds divided 1–3 times, small, tufted.
» Laminae glabrous.
» Sori without indusia.

Notogrammitis *
» Fronds undivided, small, tufted.
» Laminae with hairs or glabrous.
» Sori parallel or diagonal to midvein, without indusia.

Asplenium *
» Fronds undivided to divided five times, usually tufted.
» Laminae with scales, often sparse (except one species densely hairy); scales with prominent cell walls.
» Sori of a lamina segment diagonal to midvein, usually with indusia.

Ptisana
» Fronds large, divided twice, tufted.
» Sporangia in fused clusters just inside lamina margin.

Athyrium & Diplazium
» Fronds divided 2–4 times, tufted.
» Laminae with scales; scales without prominent cell walls.
» Sori of a lamina segment diagonal to midvein, with indusia.

Deparia
» Fronds divided 2–3 times, spread along creeping rhizomes.
» Laminae with scales and hairs; scales without prominent cell walls.
» Sori of a lamina segment diagonal to midvein, with indusia.

Blechnum *
» Fronds mostly divided once.
» Laminae with netted veins.
» Sori in short lines parallel to midvein, with indusia.

SORI CIRCULAR OR SHORT, AND AWAY FROM LAMINA MARGIN

Arthropteris

» Fronds divided once, on creeping rhizomes that usually climb.
» Laminae glabrous.
» Sori just inside lamina margins.
» Indusia absent.

Lecanopteris

» Fronds usually lobed once, spread along creeping rhizomes, often climbing.
» Laminae mostly glabrous, but rhizomes with scales.
» Indusia absent.

Polypodium

» Fronds divided once, spread along creeping rhizomes; not climbing.
» Laminae mostly glabrous, but rhizomes with scales.
» Indusia absent.

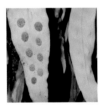

Loxogramme

» Fronds undivided, usually small, tufted, but rhizomes with proliferous roots.
» Lamina glabrous.
» Indusia absent.

Notogrammitis *

» Fronds undivided except one species divided up to two times, usually small, tufted or spread along creeping rhizomes.
» Lamina with hairs or glabrous.
» Indusia absent.

Pyrrosia

» Fronds undivided, small, spread along creeping rhizomes, usually climbing.
» Laminae with star-shaped hairs, dense and whitish on underside.
» Sori in several ill-defined rows.
» Indusia absent.

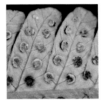

Christella

» Fronds divided once with secondary lobing, tufted although rhizomes may creep.
» Laminae with hairs.
» Basal veins of adjacent secondary lobes joining.
» Indusia kidney-shaped.

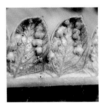

Cyclosorus

» Fronds divided once with secondary lobing, spread along creeping rhizomes.
» Laminae with scales and obscure hairs.
» Basal veins of adjacent secondary lobes joining.
» Indusia kidney-shaped.

Thelypteris

» Fronds divided twice, spread along creeping rhizomes.
» Laminae with scales and obscure hairs.
» Lamina veins free.
» Indusia kidney-shaped.

Pakau

» Fronds divided once with secondary lobing, tufted.
» Laminae seemingly glabrous.
» Basal veins of adjacent secondary lobes joining.
» Indusia absent.

SORI CIRCULAR OR SHORT, AND AWAY FROM **LAMINA MARGIN**

CONTINUED

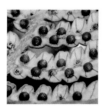

Cyathea
» Fronds large, divided 2–4 times, with scales, tufted and usually on a trunk; underside green or white.
» Indusia variable, partially or completely enveloping sori, or absent.

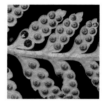

Rumohra
» Fronds divided 2–3 times, spread along creeping rhizomes, often climbing.
» Fronds with scales.
» Indusia round.

Cyrtomium
» Fronds once-divided, tufted.
» Fronds withs scales, mostly on stipe.
» Indusia round.

Cystopteris
» Fronds divided 1–3 times, small, on short-creeping rhizomes.
» Laminae mostly glabrous.
» Indusia hood-like, partially covering sori from one side.

Dryopteris
» Fronds divided 2–3 times, tufted.
» Fronds with scales.
» Indusia kidney-shaped.

Dicranopteris
» Fronds forking, mostly without lamina segments below the forks.
» Lamina segments > 4mm long.
» Many sori per lamina segment, with 7–12 sporangia per sorus.
» Indusia absent.

Lastreopsis & Parapolystichum
» Fronds divided 3–5 times, tufted to spread along creeping rhizomes.
» Fronds with hairs.
» Indusia kidney-shaped.

Sticherus
» Fronds forking, with lamina segments below most forks.
» Lamina segments > 4mm long.
» Many sori per lamina segment, with 3–5 sporangia per sorus.
» Indusia absent.

Polystichum
» Fronds divided 2–3 times, tufted.
» Fronds with scales.
» Indusia round, or absent in one species.

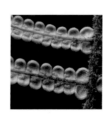

Gleichenia
» Fronds mostly forking.
» Lamina segments < 3mm long.
» One sorus per lamina segment, with 2–4 sporangia per sorus.
» Indusia absent.

SORI CIRCULAR OR SHORT, AND AWAY FROM **LAMINA MARGIN**

CONTINUED

SPORANGIA NOT IN DISCRETE SORI OF REGULAR SHAPE AND SIZE, **LACKING INDUSIA**

Hypolepis *
- » Fronds divided 3–4 times, with hairs (sometimes glandular), spread along creeping rhizomes.
- » Indusia absent.

Leptopteris
- » Fronds divided three times, tufted.
- » Laminae translucent; axes with hairs.

Leptolepia
- » Fronds divided 3–5 times, spread along creeping rhizomes.
- » Laminae with few or no hairs.
- » Sori just inside lamina margin with indusia opening outwards.

Todea
- » Fronds divided twice, tufted.
- » Laminae not transparent, mostly glabrous.

Nephrolepis
- » Fronds once-divided, tufted but rhizomes with proliferous stolons.
- » Rachis with scales.
- » Indusia shaped like a half-moon.

Osmunda
- » Fronds divided twice, tufted.
- » Laminae glabrous or hairy on axes.
- » Fertile fronds with upper pinnae dimorphic, modified into clusters of sporangia.

Platycerium
- » Fronds dimorphic, with rounded nest fronds and forking foliage fronds.
- » Sporangia at apices of foliage fronds' underside.
- » Star-shaped hairs on underside of foliage fronds.

STEMS WITH SINGLE-VEINED LEAVES

Huperzia
» Stems upright, growing on ground.
» Sporophylls not in cones, and not restricted to apex of stem.

Equisetum
» Upright, ribbed stems with whorls of branches or scale-like leaves.
» Sporangia in apical cones.

*Lycopodiella &
Lycopodium*
» Stems prostrate or upright, climbing in one species.
» Sporophylls in cones, that are lateral or at apex of stems, stalked or sessile, upright or pendulous.

Psilotum
» Stems repeatedly forking, often upright.
» Leaves inconspicuous.
» Sporangia fused in yellow clusters of three.

Phlegmariurus
» Stems generally pendulous, at least at their apex.
» Often epiphytic.
» Sporophylls at apex of stems, usually in cones.

Tmesipteris
» Stems usually epiphytic on tree fern trunks, usually pendulous.
» Sporangia fused in clusters of two near the base of a leaf.

Phylloglossum
» Plants ≤ 50mm tall.
» Sporophylls in a stalked cone surrounded by a rosette of leaves.

Selaginella
» Stems prostrate, matting ground.
» Leaves delicate.
» Sporophylls in inconspicuous green cones.

OTHER

Azolla

» Aquatic, free-floating.
» Individual leaves *c.* 1mm long.

Lygodium

» Climbing plants with forking stems.
» Fertile frond segments more divided than sterile segments.

Isoetes

» Aquatic, submerged, usually small.
» Leaves undivided, tapering cylindrical, tufted.
» Sporangia in swollen leaf bases.

Schizaea

» Fronds usually small.
» Fertile branches in apical combs.
» Sterile stems unbranched or branched.

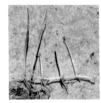

Pilularia

» Aquatic, submerged, small.
» Leaves undivided, narrowly cylindrical, spread along creeping rhizomes.

Botrychium

» Fronds with lower part divided.
» When fertile, upper part with clusters of sporangia on branched stems.

Ophioglossum

» Fronds with lower part undivided, small.
» When fertile, upper part with embedded sporangia on unbranched stem.

Anogramma leptophylla, page 72.

Sticherus flabellatus, page 254.

Species scan

Species scan

This section provides a quick aid to identification by species and should be used in conjunction with the **Species profiles.** The images are arranged by evolutionary relationships, which means that similar-looking species can be found on the same or neighbouring pages. The lycophytes are given first. The ferns are arranged by the taxonomic rank of order (with, approximately, the oldest first) and then alphabetical by family, genus and species.

LYCOPHYTES

Isoetes

Huperzia australiana

Lycopodiella cernua

Lycopodiella diffusa

Lycopodiella lateralis

Lycopodiella serpentina

Lycopodium deuterodensum

Lycopodium fastigiatum

Lycopodium scariosum

Lycopodium volubile

Phlegmariurus billardierei

Phlegmariurus varius

Phylloglossum drummondii

Selaginella kraussiana

FERNS

Equisetum arvense

Equisetum hyemale

Botrychium australe

Botrychium biforme

Ophioglossum coriaceum

Psilotum nudum

Tmesipteris elongata

Tmesipteris horomaka

Tmesipteris lanceolata

Tmesipteris sigmatifolia

Tmesipteris tannensis

Ptisana salicina

Leptopteris hymenophylloides

Leptopteris superba

Osmunda regalis

Todea barbara

Dicranopteris linearis

Gleichenia alpina

Gleichenia dicarpa

Gleichenia inclusisora

Gleichenia microphylla

Sticherus cunninghamii

Sticherus flabellatus

Hymenophyllum armstrongii

Hymenophyllum australe

Hymenophyllum bivalve

Hymenophyllum cupressiforme

Hymenophyllum demissum

Hymenophyllum dilatatum

Hymenophyllum flabellatum

Hymenophyllum flexuosum

Hymenophyllum frankliniae

Hymenophyllum lyallii

Hymenophyllum malingii

Hymenophyllum minimum

Hymenophyllum multifidum

Hymenophyllum nephrophyllum

Hymenophyllum peltatum

Hymenophyllum pluviatile

Hymenophyllum pulcherrimum

Hymenophyllum rarum

Hymenophyllum revolutum

Hymenophyllum rufescens

Hymenophyllum sanguinolentum

Hymenophyllum scabrum

Hymenophyllum villosum

Trichomanes colensoi

Trichomanes elongatum

Trichomanes endlicherianum

Trichomanes strictum

Trichomanes venosum

Lygodium articulatum

Schizaea australis

Schizaea bifida

Schizaea dichotoma

Schizaea fistulosa

Pilularia novae-hollandiae

Azolla pinnata

Azolla rubra

Cyathea colensoi

Cyathea cooperi

Cyathea cunninghamii

Cyathea dealbata

Cyathea medullaris

Cyathea smithii

Dicksonia fibrosa

Dicksonia lanata subsp. hispida

Dicksonia lanata subsp. lanata

Dicksonia squarrosa

Loxsoma cunninghamii

Asplenium aethiopicum

Asplenium appendiculatum subsp. appendiculatum

Asplenium appendiculatum subsp. maritimum

Asplenium bulbiferum

Asplenium cimmeriorum

Asplenium decurrens

Asplenium flabellifolium

Asplenium floccidum subsp. floccidum

Asplenium floccidum subsp. haurakiense

Asplenium gracillimum

Asplenium hookerianum

Asplenium hookerianum

Asplenium lamprophyllum

Asplenium lepidotum

Asplenium ×lucrosum

Asplenium lyallii

Asplenium oblongifolium

Asplenium obtusatum

Asplenium polyodon

Asplenium richardii

Asplenium scleroprium

Asplenium scolopendrium

Asplenium subglandulosum

Asplenium trichomanes

Athyrium filix-femina

Deparia petersenii subsp. congrua

Diplazium australe

Blechnum banksii

Blechnum chambersii

Blechnum colensoi

Blechnum deltoides

Blechnum discolor

Blechnum durum

Blechnum filiforme

Blechnum filiforme

Blechnum fluviatile

Blechnum fraseri

Blechnum membranaceum

Blechnum minus

Blechnum molle

Blechnum montanum

Blechnum nigrum

Blechnum norfolkianum

Blechnum novae-zelandiae

Blechnum parrisiae

Blechnum penna-marina

Blechnum procerum

Blechnum triangularifolium

Blechnum zeelandicum

Cystopteris fragilis

Cystopteris tasmanica

Histiopteris incisa

Hiya distans

Hypolepis ambigua

Hypolepis dicksonioides

Hypolepis lactea

Hypolepis millefolium

Hypolepis rufobarbata

Leptolepia novae-zelandiae

Paesia scaberula

Pteridium esculentum

Cyrtomium falcatum

Dryopteris affinis

Dryopteris dilatata

Dryopteris filix-mas

Lastreopsis hispida

Lastreopsis velutina

Parapolystichum glabellum

Parapolystichum microsorum

Polystichum cystostegia

Polystichum neozelandicum

Polystichum oculatum

Polystichum sylvaticum

Polystichum vestitum

Polystichum wawranum

Rumohra adiantiformis

Lindsaea linearis

Lindsaea trichomanoides

Lindsaea viridis

Nephrolepis cordifolia

Nephrolepis flexuosa

Lecanopteris novae-zealandiae

Lecanopteris pustulata

Lecanopteris scandens

Loxogramme dictyopteris

Notogrammitis angustifolia

Notogrammitis billardierei

Notogrammitis ciliata

Notogrammitis crassior

Notogrammitis givenii

Notogrammitis heterophylla

Notogrammitis patagonica

Notogrammitis pseudociliata

Notogrammitis rawlingsii

Platycerium bifurcatum

Polypodium vulgare

Pyrrosia elaeagnifolia

Adiantum aethiopicum

Adiantum capillus-veneris

Adiantum cunninghamii

Adiantum diaphanum

Adiantum formosum

Adiantum fulvum

Adiantum hispidulum

Adiantum raddianum

Anogramma leptophylla

Cheilanthes distans

Cheilanthes sieberi

Pellaea calidirupium

Pellaea rotundifolia

Pteris carsei

Pteris cretica

Pteris macilenta

Pteris saxatilis

Pteris tremula

Pteris vittata

Arthropteris tenella

Christella dentata

Cyclosorus interruptus

Pakau pennigera

Thelypteris confluens

Blechnum montanum, page 112.

Species profiles

Fronds

Lamina segments fan-shaped

Segments shallowly notched; indusia kidney-shaped

Adiantum aethiopicum

mākaka

PTERIDACEAE

Distinguished by its more-or-less tufted fronds divided 3–4 times, with stolons giving rise to new plants; fan-shaped lamina segments, which are generally wider than long, and none of which are incised more deeply than the indusial notches; and kidney-shaped indusia.

Grows in open, lowland, often coastal forest and scrub. Seemingly extinct in Te Waipounamu South Island.

Compare with *Adiantum capillus-veneris* and *A. raddianum*.

Frond length: 125–500mm

Indigenous to Aotearoa New Zealand, Australia, New Caledonia and Africa.

Frond divided three times

Small plant on brick wall

Young oblong indusia

Lamina segments fan-shaped; indusia oblong

Adiantum capillus-veneris[†]

Venus-hair fern

PTERIDACEAE

Distinguished by its fronds divided 1–3 times, somewhat spread along creeping rhizomes; fan-shaped lamina segments; and oblong indusia.

Grows around urban areas, commonly on brick or concrete walls, but also soil banks. Often cultivated.

Compare with *Adiantum aethiopicum* and *A. raddianum*.

Frond length 40–435mm

Naturalised in Aotearoa New Zealand. Indigenous to many other areas.

Frond, with mostly oblong lamina segments

Frond axes without hairs

Lamina underside blue-green; indusia kidney-shaped

Rhizome long-creeping

Adiantum cunninghamii

puhinui, Cunningham's maidenhair

PTERIDACEAE

Distinguished by its fronds usually divided 2–3 times, spread along creeping rhizomes; somewhat oblong lamina segments with undersides almost always glabrous and blue-green; glabrous frond axes; and kidney-shaped, glabrous indusia.

Grows in forest and scrub, particularly on soil banks. Can be abundant on mudstone and other base-rich substrates.

Compare with *Adiantum fulvum*.

Frond length 60–900mm

Endemic to Aotearoa New Zealand.

Frond divided once

Frond divided twice

Frond axes glabrous

Many plants, mostly with fronds divided once

Segment undersides and indusia can be hairy

Adiantum diaphanum

small maidenhair

PTERIDACEAE

Distinguished by its small, tufted fronds divided 1–2 times; somewhat oblong lamina segments with undersides hairy or glabrous; glabrous frond axes; and kidney-shaped indusia that can be hairy or glabrous.

Grows in lowland, warmer forests, often on soil banks.

Frond length: 50–390mm

Indigenous to Aotearoa New Zealand, Australia, many Pacific islands and south-eastern Asia.

Large, much-divided frond

Many large fronds, spread along creeping rhizomes

Rachis (at left) glabrous but costae hairy

Indusia broadly kidney-shaped

Adiantum formosum

giant maidenhair

PTERIDACEAE

Distinguished by its fronds usually divided 4 times, spread along creeping rhizomes; somewhat oblong lamina segments with undersides hairy or glabrous; glabrous stipe and rachis but hairy costae; and broadly kidney-shaped, glabrous indusia.

Grows in forest around the Manawatū Gorge. Te Tai Tokerau Northland indigenous populations are extinct. Regularly establishes itself when planted outside its native range.

Frond length: 470–970mm

Indigenous to Aotearoa New Zealand and Australia.

Lamina segments can have rounded apices

Rachis and costae hairy

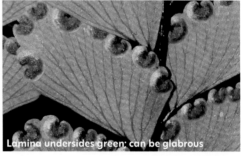

Lamina undersides green; can be glabrous

Lamina segments can have pointed apices

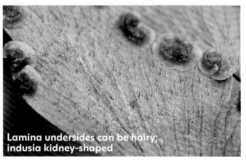

Lamina undersides can be hairy; indusia kidney-shaped

Adiantum fulvum

Adiantum viridescens

PTERIDACEAE

Distinguished by its fronds usually divided three times, spread along creeping rhizomes; somewhat oblong lamina segments with undersides hairy or glabrous; hairy frond axes; and kidney-shaped, glabrous indusia.

Grows in lowland forest and scrub.

Compare with *Adiantum cunninghamii*.

Frond length: 170–850mm

Endemic to Aotearoa New Zealand.

Fronds somewhat hand-shaped, with branching mostly basal

Young fronds can be red

Frond axes hairy; lamina segments oblong

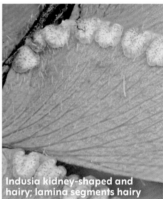

Indusia kidney-shaped and hairy; lamina segments hairy

Adiantum hispidulum

rosy maidenhair

PTERIDACEAE

Distinguished by its more-or-less tufted, somewhat hand-shaped fronds divided 2–3 times, with branching mostly at the base of the lamina, and with stolons giving rise to new plants; somewhat oblong, hairy lamina segments; hairy frond axes; and kidney-shaped, hairy indusia.

Grows in lowland forest and scrub.

Frond length: 100–730mm

Indigenous to Aotearoa New Zealand, Australia, many Pacific islands, Asia and Africa.

Frond much-divided

Segments fan-shaped, deeply incised

Indusia kidney-shaped

Adiantum raddianum[+]

delta maidenhair fern

PTERIDACEAE

Distinguished by its fronds usually divided 3–4 times, at least somewhat spread along creeping rhizomes; fan-shaped lamina segments, which are generally longer than wide, and some of which are incised more deeply than the indusial notches into two or more distinct lobes; and kidney-shaped indusia.

Commonly cultivated. Increasingly weedy in some areas.

Compare with *Adiantum aethiopicum* and *A. capillus-veneris*.

Frond length: 125–680mm

Naturalised in Aotearoa New Zealand. Indigenous to Central and South America.

Underside of fertile frond

Sori elongated along the veins, without indusia

Fronds tufted

Anogramma leptophylla

annual fern

PTERIDACEAE

Distinguished by its small, tufted fronds divided 1–3 times when fertile; glabrous lamina; and unprotected sori elongated along the veins. It is New Zealand's only annual fern, sprouting during winter.

Grows in open habitats, usually on banks, often in areas that are particularly dry over summer.

Frond length: 15–115mm

Indigenous to Aotearoa New Zealand, Australia, Asia, Africa, Europe, and Central and South America.

Adult fronds on rhizome climbing trunk of tree

Juvenile fronds on rock

Sori round, without indusia

Arthropteris tenella

jointed fern

TECTARIACEAE

Distinguished by its fronds divided once, spread along creeping and climbing rhizomes; pinnae jointed to the rachis; and round sori without indusia close to the pinna margins.

Grows in lowland forest, often on rocky ground. Juveniles frequently scramble over rocks and roots before climbing a tree trunk to form fertile fronds.

Frond length: 85–560mm

Indigenous to Aotearoa
New Zealand and Australia

Frond divided 2-3 times

Sori elongated

Stipes somewhat tufted

Rachis densely scaly

Asplenium aethiopicum[+]

ASPLENIACEAE

Distinguished by its more-or-less tufted fronds divided 2–3 times; densely scaly rachis that is largely dark brown; and long sori.

Grows on rocks, walls, concrete and the ground, mostly within Tāmaki Makaurau Auckland.

Compare with *Asplenium polyodon*.

Frond length: 110–630mm

Naturalised in Aotearoa New Zealand. Indigenous to Africa, Asia, and Australia.

Frond partially upright, with triangular primary pinnae and many tertiary segments

Larger frond with drooping apex

Frond underside with sori

Asplenium appendiculatum
subsp. *appendiculatum*

Asplenium terrestre subsp. *terrestre*
ground spleenwort

ASPLENIACEAE

Distinguished by its tufted, at least partially upright, leathery fronds that are divided 2–3 times, with triangular primary pinnae that have markedly longer secondary segments towards their base; and green rachis upperside. The lamina segments are narrow.

Grows on the ground in drier forests, mostly away from the coast.

Compare with subsp. *maritimum* and *Asplenium flaccidum*.

Frond length: 80–640mm

Indigenous to Aotearoa
New Zealand and Australia.

Large frond, with many tertiary segments

Lamina segments broad

Fronds upright; segments fleshy

Frond underside with sori

Asplenium appendiculatum subsp. *maritimum*

Asplenium terrestre subsp. *maritimum*

ASPLENIACEAE

Distinguished by its tufted, upright, leathery or fleshy fronds divided 2–3 times, with rectangular primary pinnae that have only slightly longer secondary segments towards their base; and green rachis upperside. The lamina segments are often broad.

Grows only near the coast, on the ground in forest and scrub, and on exposed rocks.

Compare with *Asplenium appendiculatum* subsp. *appendiculatum*.

Frond length: 40–670mm

Endemic to Aotearoa New Zealand.

Frond large, somewhat rectangular, with closely set pinnae

Frond divided three times, with many bulbils

Rachis scales broad

Sori

Asplenium bulbiferum

mouku, hen and chickens fern

ASPLENIACEAE

Distinguished by its more-or-less tufted, large, rectangular fronds divided 2–3 times, with closely set, broad pinnae; mostly green rachis upperside, with lateral wings; broad rachis scales mostly without hairlike apices; and many bulbils on older fronds. Fronds < 300mm long are usually sterile.

Grows in forest, usually in wetter habitats.

Compare with *Asplenium gracillimum* and *A. ×lucrosum*.

Frond length: 290–1600mm

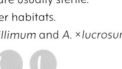

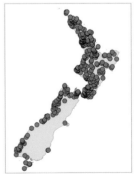

Endemic to Aotearoa New Zealand.

Relatively large plant among limestone rocks

Small, triangular frond

Frond underside with sori; rachis scales narrow

Asplenium cimmeriorum

cave spleenwort

ASPLENIACEAE

Distinguished by its small, somewhat rectangular or triangular fronds divided 2–3 times, with broad pinnae; green rachis upperside; narrow rachis scales; and lack of bulbils. The rhizome is shortly creeping, but the fronds are often tufted. Fronds > 50mm long are usually fertile.

Grows on limestone, often within or around caves, in western Waikato and north-western Te Waipounamu South Island.

Compare with *Asplenium gracillimum* and *A. hookerianum*.

Frond length: 45–370mm

Endemic to Aotearoa New Zealand.

Fronds divided once

Sori

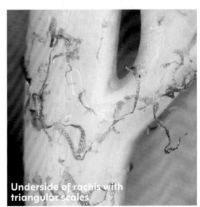

Underside of rachis with triangular scales

Asplenium decurrens

Asplenium northlandicum, A. obtusatum
  subsp. *northlandicum*

ASPLENIACEAE

Distinguished by its tufted, often fleshy fronds divided once; mostly green rachis upperside; pinnae with rounded apices; and triangular rachis scales.

Grows only near northern coasts, on exposed rocks and under scrub.

Compare with *Asplenium lepidotum, A. oblongifolium* and *A. obtusatum.*

Frond length: 35–790mm

Indigenous to Aotearoa New Zealand and Australia, and probably some Pacific islands.

Fronds divided once

Rachis green; pinnae fan-shaped

Rachis apex rooting, with new plant

Asplenium flabellifolium

necklace fern

ASPLENIACEAE

Distinguished by its small (usually < 300mm long), mostly prostrate, narrow fronds divided once, with fan-shaped pinnae; and green rachis. While the fronds are tufted, the rachises often root at their extended apex to make new plants, so that there can be many entangled plants matting the ground.

Grows on the ground in drier forest or scrub, or in the open.

Compare with *Asplenium trichomanes* and *Lindsaea linearis*.

Frond length: 45–550mm

Indigenous to Aotearoa
New Zealand and Australia.

Fronds divided twice

Fronds pendulous with rectangular primary pinnae

Frond underside with elongated sori

Asplenium flaccidum subsp. *flaccidum*

makawe, hanging spleenwort

ASPLENIACEAE

Distinguished by its tufted, pendulous, leathery fronds usually divided twice, with somewhat rectangular primary pinnae that usually have secondary segments all about equal in length; and mostly green rachis upperside.

Grows in forest and scrub, usually as an epiphyte, but also on the ground, particularly on dry soils or under pines.

Compare with *Asplenium appendiculatum*.

Frond length: 75–1560mm

Indigenous to Aotearoa New Zealand and Australia.

Large plant with partially upright fronds.

Smaller plant with upright fronds

Frond with some tertiary segments

Secondary segments next to rachis and facing apex are longer

Asplenium flaccidum subsp. *haurakiense*

Asplenium haurakiense

ASPLENIACEAE

Distinguished by its tufted, partially upright, leathery fronds divided 2–3 times, with the secondary segments next to the rachis and facing the frond apex usually longer than the others; and green rachis upperside.

Grows only near the coast, usually on the ground but also as a low epiphyte on pōhutukawa.

Compare with *Asplenium flaccidum* subsp. *flaccidum*.

Frond length: 45–680mm

Endemic to Aotearoa New Zealand.

Frond diamond-shaped and divided three times

Frond segments rarely very narrow

Sori

Scales narrow

Asplenium gracillimum

Asplenium bulbiferum subsp. *gracillimum*

ASPLENIACEAE

Distinguished by its tufted, medium-sized, somewhat diamond-shaped fronds divided 2–4 times, with widely spaced pinnae that are usually broad but rarely narrow; mostly green rachis upperside, without broad lateral wings; narrow rachis scales with hairlike apices; and few or no bulbils. Fronds > 200mm long are usually fertile.

Grows in forest and scrub, usually in somewhat drier habitats.

Compare with *Asplenium bulbiferum*, with which it often hybridises.

Frond length: 105–1080mm

Indigenous to Aotearoa New Zealand and Australia.

Broad lamina segments

Intermediate lamina segments

Small plants

Narrow lamina segments

Frond underside with sori; rachis scales narrow

Asplenium hookerianum

Asplenium colensoi
Hooker's spleenwort

ASPLENIACEAE

Distinguished by its tufted, small, somewhat diamond-shaped fronds usually divided 2–4 times, with widely spaced lamina segments that can be broad or narrow; green rachis upperside; narrow rachis scales; and lack of bulbils. The primary pinnae do not arch upwards. Fronds > 50mm long are usually fertile.

Grows on the ground in forest and scrub, mostly lowland or montane, and rarely above 1000m elevation.

Compare with *Asplenium gracillimum* and *A. richardii.*

Frond length: 25–425mm

Indigenous to Aotearoa
New Zealand and Australia.

Frond with glossy upperside and closely set secondary segments

Stipes arising from creeping rhizome

Frond underside, with long sori

Asplenium lamprophyllum

ASPLENIACEAE

Distinguished by its fronds divided 2–3 times, with closely set, broad secondary pinnae; green rachis upperside; and lack of bulbils. The rhizome is creeping although the fronds can be more or less tufted. The fronds are glossy and can smell of oil of wintergreen, and the sori are long and curving.

Grows in warmer forest, usually on the ground but rarely as a low epiphyte.

Frond length: 100–820mm

Endemic to Aotearoa New Zealand.

Frond divided once

Upperside of young frond dotted with small scales

Underside of frond with triangular scales

Asplenium lepidotum

ASPLENIACEAE

Distinguished by its tufted fronds divided once (or, rarely, undivided); mostly green rachis upperside; and narrowly triangular rachis scales. The uppersides of young fronds are abundantly scaly.

Grows on limestone and other base-rich substrates, in forest, scrub or on exposed rocks.

Compare with *Asplenium lyallii*, *A. oblongifolium* and *A. obtusatum*.

Frond length: 120–1100mm

Endemic to Aotearoa New Zealand.

Dimorphism, with narrow fertile segments at top

Frond, dimorphic and with bulbils

Bulbils growing from frond upperside

Asplenium ×*lucrosum*⁺

false hen and chickens fern

ASPLENIACEAE

Distinguished by its tufted fronds divided 2–4 times; dimorphism with narrow fertile frond segments and broad sterile frond segments; mostly green rachis upperside; and many bulbils on upperside of mature fronds.

Grows mostly in gardens, where common. The map shows records of vegetative propagation around planted individuals.

Compare with *Asplenium bulbiferum*. Sterile hybrid between New Zealand's *Asplenium bulbiferum* and Norfolk Island's *A. dimorphum*.

Frond length: 470–1200mm

Naturalised in Aotearoa New Zealand. Hybrid of horticultural origin.

Frond with many secondary segments

Frond with few secondary segments

Lamina underside with elongated sori

Rachis scales

Asplenium lyallii

Lyall's spleenwort

ASPLENIACEAE

Distinguished by its tufted fronds divided 1–2 times, with broad pinnae; green rachis upperside; and narrowly triangular rachis scales. The upperside of young fronds sometimes has many scales. The apical frond segment is usually lobed even in once-divided fronds.

Grows on limestone and other base-rich substrates, in forest, scrub or on exposed rocks.

Compare with *Asplenium lepidotum*.

Frond length: 12–1160mm

Endemic to Aotearoa New Zealand.

Fronds divided once; glossy upperside

Frond underside with elongated sori

Scales narrow and hairlike

Asplenium oblongifolium

huruhuruwhenua, shining spleenwort

ASPLENIACEAE

Distinguished by its more-or-less tufted fronds divided once; glossy lamina upperside; mostly green rachis upperside; and hairlike rachis scales.

Grows in forest and scrub, usually below 400m elevation.

Compare with *Asplenium decurrens*, *A. lepidotum* and *A. obtusatum*.

Frond length: 70–1400mm

Endemic to Aotearoa New Zealand.

Pinnae with pointed apices

Pinnae with rounded apices

Frond underside, with elongated sori

Scales broadly triangular

Asplenium obtusatum

paranako, shore spleenwort

ASPLENIACEAE

Distinguished by its more-or-less tufted, often fleshy fronds divided once; mostly green rachis upperside; and broadly triangular rachis scales. The pinna apices can be pointed or rounded.

Grows only near southern coasts, on the ground in forest and scrub, and on exposed rocks.

Compare with *Asplenium decurrens, A. lepidotum* and *A. oblongifolium.*

Frond length: 60–1000mm

Indigenous to Aotearoa New Zealand, Southern Ocean islands and maybe South America.

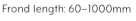

Frond underside with dark-brown rachis

Fronds divided once, with prominently toothed pinnae

Lamina underside with long, curving sori

Asplenium polyodon

petako, sickle spleenwort

ASPLENIACEAE

Distinguished by its more-or-less tufted fronds divided once, with pinnae that have prominently toothed margins; and dark-brown rachis. The sori are long and curving.

Grows in forest and scrub, usually at lower elevations, and usually as an epiphyte but also on the ground.

Frond length: 115–1640mm

Indigenous to Aotearoa New Zealand, Australia and New Caledonia.

Frond with overlapping, arching primary pinnae

Rachis underside with narrow scales

Asplenium richardii

matua-kaponga, Richard's spleenwort

ASPLENIACEAE

Distinguished by its tufted, somewhat diamond-shaped fronds divided 2–4 times, with overlapping, narrow segments; green rachis upperside; narrow rachis scales; and lack of bulbils. The primary pinnae arch upwards from the rachis, making the frond three-dimensional.

Grows on the ground in drier forest and scrub, usually above 600m elevation.

Compare with *Asplenium appendiculatum* and *A. hookerianum*.

Frond length: 40–470mm

Endemic to Aotearoa New Zealand.

Frond divided once

Lamina margins deeply toothed

Sori

Asplenium scleroprium

ASPLENIACEAE

Distinguished by its tufted, leathery or fleshy fronds divided once, with deeply toothed margins; and mostly green rachis upperside. The sori can reach the margin at the indentations.

Grows near far-southern coasts, on the ground in forest and scrub.

Compare with *Asplenium obtusatum*.

Frond length: 170–1060mm

Endemic to Aotearoa New Zealand.

Undivided fronds with elongated sori

Asplenium scolopendrium[+]

hart's tongue fern

ASPLENIACEAE

Distinguished by its tufted, undivided fronds, with the base of the lamina having lobes that clasp the stipe; and sori paired along the side veins.

Grows on the ground or on rock walls, usually near urban areas.

Frond length: 140–500mm

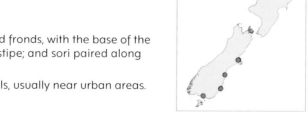

Naturalised in Aotearoa New Zealand. Indigenous to Europe and Asia.

Plant in rock crevice

Frond upperside hairy

Frond underside hairy

Asplenium subglandulosum

Pleurosorus rutifolius
blanket fern

ASPLENIACEAE

Distinguished by its tufted, small (≤ 190mm long) fronds divided 1–2 times, with a dense cover of hairs; green rachis upperside; and sori lacking indusia.

Grows on or among rocks, in the open or within scrub, in drier regions.

Frond length: 25–190mm

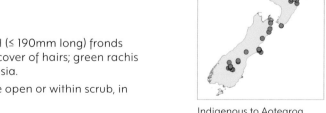

Indigenous to Aotearoa
New Zealand and Australia.

Plants growing on limestone

Fronds divided once

Rachis dark-brown; sori elongated

Asplenium trichomanes

maidenhair spleenwort

ASPLENIACEAE

Distinguished by its tufted, small (≤ 300mm long), narrow fronds divided once with many oblong pinnae; and dark-brown rachis.

Grows usually on limestone or other base-rich substrates, in the open or within forest and scrub.

Compare with *Asplenium flabellifolium*.

Frond length: 25–300mm

Indigenous to Aotearoa New Zealand and many parts of the world.

Frond divided three times

Grooves of axes V-shaped

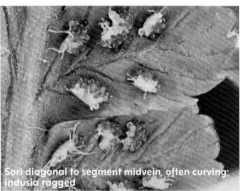
Sori diagonal to segment midvein, often curving; indusia ragged

Stipes tufted, scaly

Athyrium filix-femina[+]

lady fern

ATHYRIACEAE

Distinguished by its tufted, scaly but not hairy fronds divided 2–3 times; grooves on the upperside of the rachis and costae V-shaped, connected or not; and sori often curved with indusia margins deeply toothed or ragged.

Grows in disturbed areas near settlements, often near waterways.

Compare with *Diplazium australe*.

Frond length: 260–950mm

Naturalised in Aotearoa New Zealand. Indigenous to many regions of the northern hemisphere.

Triangular frond with regular branching

With (at top) *Azolla rubra*

Matting surface of water

Red from growing in the open

Azolla pinnata[+]

ferny azolla

SALVINIACEAE

Distinguished by its aquatic habitat; free-floating fronds with regular branching; and roots with fine rootlets. Becomes red in the open.

Grows on slow-moving water surfaces, where it can form extensive mats. Is displacing the indigenous *Azolla rubra* in the north.

Frond length: 5–25mm

Naturalised in Aotearoa New Zealand. Indigenous to Africa, Asia, Australia and New Caledonia.

Frond with irregular branching

Irregularly shaped fronds

A. pinnata, roots with rootlets (left); *A. rubra*, without rootlets (right)

Matting pond surface

Azolla rubra

retoreto, Pacific azolla

SALVINIACEAE

Distinguished by its aquatic habitat; free-floating fronds with irregular branching; and roots without fine rootlets. Becomes red in the open.

Grows on slow-moving water surfaces, where it can form extensive mats.

Previously known as *Azolla filiculoides*, but that species is confined to the Americas.

Frond length: 5–40mm

Indigenous to Aotearoa New Zealand, Australia, Indonesia and Japan.

Fronds tufted

Rhizome scales uniformly brown

Fleshy sterile fronds with sessile pinnae, and shorter fertile fronds

Blechnum banksii

Austroblechnum banksii
shore hard fern

BLECHNACEAE

Distinguished by its strongly dimorphic, tufted fronds divided once; fleshy, narrow sterile fronds with sessile pinnae; and uniformly brown rhizome scales. The fertile fronds are shorter than the sterile fronds.

Grows at the coast, on exposed rocks or among other vegetation.

Compare with *Blechnum chambersii*, *B. durum* and *B. membranaceum*. Previously known as *B. blechnoides*.

Frond length: 25–500mm

Indigenous to Aotearoa New Zealand and possibly Chile.

Basal pinnae alternate or nearly so

Fronds tufted, dimorphic

Frond curved, pinnae without stalks

Blechnum chambersii

Austroblechnum lanceolatum
rereti, lance fern

BLECHNACEAE

Distinguished by its strongly dimorphic, tufted fronds divided once; sterile fronds 15–115mm wide, often curving, with sessile pinnae that are much longer than wide; and basal pinnae that are usually mostly alternate or nearly so. The fertile fronds are of similar length to or shorter than the sterile fronds.

Grows in forest, where it is often common.

Compare with the smaller *Blechnum membranaceum* and the larger *B. norfolkianum*.

Frond length: 85–700mm

Indigenous to Aotearoa
New Zealand and
Australia.

Dimorphic sterile and fertile fronds; pinnae sessile

Pinnae reduced to flanges at base of frond

Blechnum colensoi

Austroblechnum colensoi
peretao, Colenso's hard fern

BLECHNACEAE

Distinguished by its strongly dimorphic fronds divided once, somewhat spread along shortly creeping rhizomes; and broad, sessile, dark-green, sterile pinnae, which become reduced to small flanges at the frond base.

Grows usually in damp and shaded forest, often on steep banks.

Frond length: 225–1270mm

Endemic to Aotearoa
New Zealand.

Pinnae sessile; basal pinnae reflexed

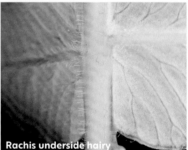

Rachis underside hairy

Dimorphic sterile and fertile fronds

Blechnum deltoides

Cranfillia deltoides
korokio

BLECHNACEAE

Distinguished by its strongly dimorphic fronds divided once, tufted or somewhat spread along shortly creeping rhizomes; and sessile pinnae, with the basal-most pair reflexed backwards and not significantly shorter than those in the middle of the frond. The rachis underside is usually covered in fine hairs.

Grows in forest or in the open, usually on banks.

Previously known as *Blechnum vulcanicum*, but that species is confined to south-eastern Asia.

Frond length: 130–1010mm

Indigenous to Aotearoa
New Zealand and Australia.

Fronds tufted

Sometimes with a short trunk

Underside of sterile fronds paler than upperside

Pinnae from middle of fertile frond

Blechnum discolor

Lomaria discolor
piupiu, crown fern

BLECHNACEAE

Distinguished by its strongly dimorphic fronds divided once, tufted and sometimes on a short trunk; sterile frond underside paler than upperside; and fertile fronds transitioning from longer fertile pinnae at their apex to sterile pinnae at their base.

Grows in forest and scrub, sometimes dominating the groundcover.

Frond length: 140–1400mm

Endemic to Aotearoa
New Zealand.

Tufted, fleshy fronds; pinnae sessile; fertile fronds dimorphic

Rhizome scales with partial black margins

Blechnum durum (left) with the generally smaller *Blechnum banksii* (right)

Blechnum durum

Austroblechnum durum

BLECHNACEAE

Distinguished by its strongly dimorphic, tufted fronds divided once; sterile fronds fleshy and broad, with sessile pinnae; and brown rhizome scales with irregular black margins. The fertile fronds are usually shorter than the sterile fronds.

Grows at the coast in the far south, on exposed rocks or among other vegetation.

Compare with *Blechnum banksii*.

Frond length: 110–820mm

Endemic to Aotearoa New Zealand.

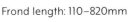

Adult sterile and fertile fronds in canopy

Adult fronds on rhizome climbing tree trunk

Juvenile fronds with pinnae margins prominently toothed

Juvenile fronds on climbing rhizome

Blechnum filiforme

Icarus filiformis
pānako, thread fern

BLECHNACEAE

Distinguished by its trimorphic (juvenile sterile, adult sterile, and adult fertile) fronds divided once, spread along creeping and climbing rhizomes; and fertile fronds with very narrow pinnae. Juvenile sterile fronds are readily recognisable by their prominently toothed lamina margins.

Grows in lowland forest. Juvenile plants can creep extensively over the ground before climbing a tree.

Frond length: 170–760mm

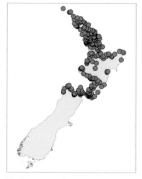

Endemic to Aotearoa New Zealand.

Tufted fronds

Dimorphic fronds

Rachis prominently scaly

Fertile fronds held higher than sterile fronds

Blechnum fluviatile

Cranfillia fluviatilis
kiwikiwi, creek fern

BLECHNACEAE

Distinguished by its strongly dimorphic, tufted, narrow fronds divided once; and abundant scales on the stipes and rachises. The sterile fronds form a rosette below the upright fertile fronds.

Grows in forests, where it is not limited to creeks.

Compare with *Pellaea rotundifolia*.

Frond length: 80–900mm

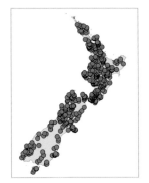

Indigenous to Aotearoa New Zealand, Australia, New Guinea and south-east Asia.

Fertile (left) and sterile fronds, only weakly dimorphic

Fronds tufted, often on a short trunk

Fronds divided twice, with jagged wing along rachis

Underside of fertile frond with immature sori

Blechnum fraseri

Diploblechnum fraseri

BLECHNACEAE

Distinguished by its weakly dimorphic fronds divided twice, which are tufted and often on a short trunk; and jagged wing along the rachis.

Grows in forest, especially kauri forest, and scrub, often in extensive colonies.

Frond length: 190–660mm

Indigenous to Aotearoa New Zealand, New Guinea and south-eastern Asia.

Fronds tufted; sterile pinnae sessile and only slightly longer than wide

Pinnae at base of frond mostly opposite

Pinnae of fertile fronds

Blechnum membranaceum

Austroblechnum membranaceum

BLECHNACEAE

Distinguished by its strongly dimorphic, tufted fronds divided once; sterile fronds 8–40mm wide, with sessile pinnae that are only slightly longer than wide; and basal pinnae that are mostly opposite. The fertile fronds are usually longer than the sterile fronds.

Grows in forests, often in dark and humid habitats. Usually less common than *B. chambersii*.

Compare with *Blechnum banksii* and *B. chambersii*. It frequently hybridises with the latter, producing intermediates.

Frond length: 45–370mm

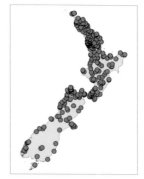

Endemic to Aotearoa New Zealand.

Medium-sized individual; dimorphic fronds

Small but still fertile plants

Fronds with shortened pinnae at base

Pinnae with short stalks; scales uniformly pale brown

Blechnum minus

Parablechnum minus
swamp kiokio

BLECHNACEAE

Distinguished by its strongly dimorphic fronds divided once, tufted or spread along creeping rhizomes; sterile fronds with shortly stalked pinnae, and much shorter pinnae at frond base; and pale-brown scales on the frond underside.

Grows usually in wet habitats, such as wetlands and alongside streams.

Compare with *Blechnum novae-zelandiae*.

Frond length: 45–1700mm

Indigenous to Aotearoa
New Zealand and Australia.

Fertile fronds with narrower pinnae

Rows of shortly rectangular sori parallel to midvein

Rachis hairy; pinnae stalked at mid-rachis

Fronds tufted

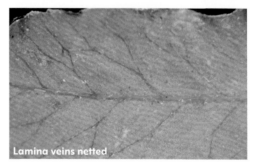

Lamina veins netted

Blechnum molle

Doodia mollis
mokimoki

BLECHNACEAE

Distinguished by its somewhat dimorphic and tufted fronds divided once (rarely twice); shortly rectangular sori in rows parallel to midvein; netted lamina veins; rachis with hairs and few scales; pinnae with short stalks extending up to at least the mid-rachis; and an apical pinna less than 1/8 of frond length.

Grows in warmer, lowland habitats; often coastal.

Compare with *Blechnum parrisiae* and *B. zeelandicum*.

Frond length: 70–620mm

Endemic to Aotearoa New Zealand.

Sterile and fertile fronds

Base of frond without markedly shortened pinnae

Shortly stalked pinnae; black-spot scales

Black-spot scales

Blechnum montanum

Parablechnum montanum
mountain kiokio

BLECHNACEAE

Distinguished by its strongly dimorphic fronds divided once, somewhat spread along creeping rhizomes; sterile fronds with shortly stalked pinnae, and lacking much-shortened pinnae at frond base; and 'black-spot' scales on the frond underside. Exposed fronds are usually olive- or orange-green.

Grows in colder, usually upland habitats, at forest margins or in subalpine scrub and tussock-land.

Compare with *Blechnum novae-zelandiae* and *B. triangularifolium*.

Frond length: 185–1150mm

Endemic to Aotearoa New Zealand.

Small, dimorphic fronds

Blechnum nigrum

Cranfillia nigra
black hard fern

BLECHNACEAE

Distinguished by its small, strongly dimorphic, tufted fronds divided once; and blackish-green sterile fronds with sessile pinnae and enlarged apical pinna.

Grows in dark, wet forest.

Frond length: 50–310mm

Endemic to Aotearoa
New Zealand.

Broad sterile fronds

With fertile fronds above

Blechnum norfolkianum

Austroblechnum norfolkianum

BLECHNACEAE

Distinguished by its strongly dimorphic, tufted fronds divided once; sterile fronds 70–250mm wide, with sessile pinnae that are much longer than wide; and basal pinnae that are usually mostly alternate or nearly so. The fertile fronds are usually a little shorter than the sterile fronds.

Grows in forest, usually on smaller islands.

Compare with the generally smaller *Blechnum chambersii*.

Frond length: 230–800mm

Indigenous to Aotearoa
New Zealand, Norfolk Island
and Vanuatu.

Fertile frond among sterile fronds

Black-spot scales on stipe

Shortly stalked pinna with black-spot scales on underside

Shortened pinnae at base of frond

Blechnum novae-zelandiae

Parablechnum novae-zelandiae
kiokio

BLECHNACEAE

Distinguished by its strongly dimorphic fronds divided once, somewhat spread along creeping rhizomes; sterile fronds with shortly stalked pinnae, and much shorter pinnae at frond base; and a basal black-spot on the otherwise brown scales on the frond underside. Fronds that are partly fertile and partly sterile are not unusual.

Grows in forest or in the open; often especially prominent on banks.

Compare with *Blechnum minus, B. montanum, B. procerum* and *B. triangularifolium.*

Frond length: 185–2780mm

Endemic to Aotearoa
New Zealand.

Pinnae mostly sessile

Reddish young fronds; fertile and sterile fronds similar

Multiple rows of shortly rectangular sori parallel to midvein

lamina veins netted

Blechnum parrisiae

Doodia australis, Doodia media subsp. *australis*
pukupuku, rasp fern

BLECHNACEAE

Distinguished by its weakly dimorphic fronds divided once, tufted or spread along creeping rhizomes; shortly rectangular sori in rows parallel to midvein; netted lamina veins; rachis with hairs and scales; and pinnae with short stalks confined to basal third of rachis.

Grows usually in warmer, lowland forest and scrub; often coastal.

Compare with *Blechnum molle* and *B. zeelandicum*.

Frond length: 130–760mm

Indigenous to Aotearoa
New Zealand and Australia.

Small, dimorphic fronds

Blechnum penna-marina

Austroblechnum penna-marina
alpine hard fern

BLECHNACEAE

Distinguished by its small, strongly dimorphic fronds divided once, spread along creeping rhizomes; and sterile fronds with sessile pinnae. The usually upright fertile fronds are longer than the sterile fronds.

Grows in open habitats and is more common at higher elevations or in colder habitats.

Frond length: 15–430mm

Indigenous to Aotearoa New Zealand, Australia and South America.

Fertile fronds upright above sterile fronds

Pinnae at base of frond not markedly shortened

Pinnae shortly stalked; scales dark-brown

Blechnum procerum

Parablechnum procerum
small kiokio

BLECHNACEAE

Distinguished by its strongly dimorphic fronds divided once, somewhat spread along creeping rhizomes; often olive-green sterile fronds with shortly stalked pinnae, and lacking much-shortened pinnae at frond base; and dark-brown scales on the frond underside. The longer, upright fertile fronds are characteristic.

Grows generally in drier or colder forest, often on poorer soils.

Compare with *Blechnum montanum* and *B. novae-zelandiae*.

Frond length: 70–1430mm

Endemic to Aotearoa
New Zealand.

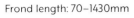

Sterile fronds

Pinnae at base of frond not markedly shortened

Scales with large black spot

Blechnum triangularifolium

Green Bay kiokio

BLECHNACEAE

Distinguished by its strongly dimorphic fronds divided once, tufted or spread along shortly creeping rhizomes; sterile fronds with shortly stalked and often curving pinnae, and lacking much-shortened pinnae at frond base; and 'black-spot' scales on the frond underside.

Grows usually on base-rich substrates such as limestone and mudstone, generally at lower elevations.

Compare with *Blechnum montanum* and *B. novae-zelandiae*.

Frond length: 200–1920mm

Endemic to Aotearoa New Zealand.

Fertile frond with very long apical pinna

Fertile fronds with narrower pinnae

Rectangular sori parallel to pinna midvein

Rachis without hairs; pinnae stalked at mid-rachis

Blechnum zeelandicum

Doodia squarrosa

BLECHNACEAE

Distinguished by its somewhat dimorphic and tufted fronds divided once; shortly rectangular sori in rows parallel to midvein; netted lamina veins; rachis with scales but no hairs; pinnae with short stalks extending up to the mid-rachis; and an elongated apical pinna at least ⅛ of frond length.

Grows in warmer, lowland forest and scrub; often coastal.

Compare with *Blechnum molle* and *B. parrisiae*.

Frond length: 85–530mm

Endemic to Aotearoa
New Zealand.

Frond with broad sterile segments; upright fertile part absent

Upright fertile part of frond present

Fronds with bronze tinge

Branched fertile part of frond (young)

Botrychium australe

Sceptridium australe
pātōtara, parsley fern

OPHIOGLOSSACEAE

Distinguished by the leafy lower part of its frond divided 3–5 times, with free veins and broad segments 1.3–5mm wide; and the fertile upright part of the frond branched with sporangia borne on the upper axes. Fronds can be green or bronze.

Grows in scrub or forest, often in disturbed areas and often on flat ground.

Frond length: 50–600mm

Indigenous to Aotearoa New Zealand, Australia and South America.

Frond with very narrow sterile segments; upright fertile part absent

Sterile and upright fertile frond parts

Frond with narrow sterile segments and bronze tinge

Branched fertile part of frond

Botrychium biforme

Sceptridium biforme
fine-leaved parsley fern

OPHIOGLOSSACEAE

Distinguished by the leafy lower part of its frond divided 5–7 times, with free veins and narrow segments 0.2–1mm wide; and the fertile upright part of the frond branched with sporangia borne on the upper axes. Fronds can be green or bronze.

Grows in forest or scrub, often on flat ground.

Frond length: 70–440mm

Endemic to Aotearoa New Zealand.

Frond divided twice

Frond underside densely scaly; sori on lamina margins

Plant on open rock

Cheilanthes distans

woolly cloak fern

PTERIDACEAE

Distinguished by its narrow fronds divided 2–3 times, somewhat spread along shortly creeping rhizomes; sori on lamina margins; and dense covering of scales and hairs.

Grows in dry, open, rocky lowland areas.

Frond length: 25–400mm

Indigenous to Aotearoa New Zealand, Australia and New Caledonia.

Frond underside with few scales and hairs

Mature sori bursting out from under inrolled margins

Plant in the open

Cheilanthes sieberi

rock fern

PTERIDACEAE

Distinguished by its narrow fronds divided 2–4 times, somewhat spread along shortly creeping rhizomes; sori on lamina margins; and by having only a few scales and hairs on its fronds.

Grows in dry, open, rocky areas.

Frond length: 60–590mm

Indigenous to Aotearoa New Zealand, Australia and New Caledonia.

Frond divided once but with deep secondary lobes

Shortened pinnae at base of frond

Frond upperside hairy, especially axes

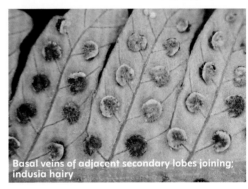

Basal veins of adjacent secondary lobes joining; indusia hairy

Christella dentata

soft fern

THELYPTERIDACEAE

Distinguished by its softly hairy fronds, divided once but with deep secondary lobes, and somewhat tufted at the apex of a creeping or erect rhizome; much shortened pinnae at base of frond; basal veins of adjacent secondary lobes joining; and sori with hairy, kidney-shaped indusia.

Grows at margins of forest and scrub, including in geothermal areas.

Compare with *Cyclosorus interruptus* and *Thelypteris confluens*.

Frond length: 340–1195mm

Indigenous to Aotearoa New Zealand and tropical and subtropical regions from Africa through Asia to the Pacific.

Fertile adult plant with no trunk

Lamina upperside dull

Stipe scales pale brown

Sori with hairs but lacking indusia

Cyathea colensoi

Alsophila colensoi
mountain tree fern

CYATHEACEAE

Distinguished by its large, tufted fronds usually divided three times, and absence of trunk; stipes with pale-brown scales; and sori away from lamina margin, with hairs but lacking indusia. A trunk-less, scaly tree fern with sori will almost certainly be *Cyathea colensoi*. The dull lamina upperside aids identification.

Grows in colder forest, especially at higher elevations.

Frond length: 460–1500mm

Endemic to Aotearoa
New Zealand.

Stipes thick, green or brown

Stipe scales pale

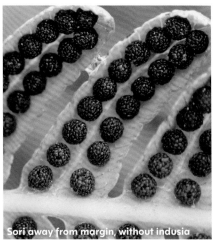

Sori away from margin, without indusia

Cyathea cooperi[†]

Sphaeropteris cooperi

CYATHEACEAE

Distinguished by its large fronds usually divided three times, borne on a trunk; green or brown stipes with pale scales; lamina scales with spines on their margins; and sori away from lamina margin, without indusia.

Grows near urban areas and increasingly weedy, particularly around Tāmaki Makaurau Auckland.

Compare with *Cyathea medullaris*.

Frond length: ?–4000mm

Naturalised in Aotearoa New Zealand. Indigenous to Australia.

Stipes narrow, black

Scales brown rather than blackish

Stipe with rough base, among brown scales

Sori with hood-shaped indusia

Cyathea cunninghamii

Alsophila cunninghamii
pūnui, slender tree fern

CYATHEACEAE

Distinguished by its large fronds divided 3–4 times, borne on a trunk; slender, blackish stipes with rough bases and brown scales; and sori away from lamina margin, with hood-shaped indusia forming more than half a sphere. Plants to c. 2m tall can have a weakly developed skirt of dead stipes and can be difficult to separate from juvenile *Cyathea smithii* – see the eFloraNZ.

Grows mostly in wetter forests, often near waterways.

Compare with *Cyathea medullaris* and *C. smithii*.

Frond length: 1500–3000mm

Indigenous to Aotearoa New Zealand and Australia.

Lamina upperside glossy

Lamina underside white

Trunk with peg-like remains of stipe bases

Stipe with white coating, warty and scaly

Sori in cup-like indusia; frond axes with curly hairs

Cyathea dealbata

Alsophila tricolor
ponga, silver fern

CYATHEACEAE

Distinguished by its large fronds usually divided 3 times, with white lamina undersides, borne on a trunk; stipes scaly and with whitish coating; curly hairs on the axes; and sori away from lamina margin, with cup-shaped indusia. The white colouring develops only in older plants (fronds > c. 1m), and is less pronounced in northern plants, which can have a silver or grey-green lamina underside.

Grows mostly in warmer, drier forest and scrub.

Frond length: 2000–4000mm

Endemic to Aotearoa
New Zealand.

Stipes thick, black; trunk with hexagonal scars

Fronds large and arching

Stipe with smooth base and dark-brown scales

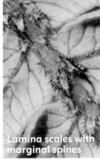

Lamina scales with marginal spines

Sori enclosed in indusia that rupture irregularly

Cyathea medullaris

Sphaeropteris medullaris
mamaku

CYATHEACEAE

Distinguished by its large fronds divided 3–4 times, borne on a tall trunk often with hexagonal stipe scars; thick, black stipes with smooth bases and dark-brown scales; lamina scales with spines on their margins; and sori away from lamina margin, with irregularly rupturing spherical indusia.

Grows mostly in warmer, wetter forests, usually at lower elevations, where it can be abundant on slips and other disturbed areas.

Compare with *Cyathea cunninghamii*.

Frond length: 3000–5000mm

Indigenous to Aotearoa New Zealand and some Pacific islands.

Skirt of dead frond stipes and rachises

Stipe scales orange-brown

Sori with saucer-shaped indusia

Cyathea smithii

Alsophila smithii
kātote, Smith's tree fern

CYATHEACEAE

Distinguished by its large fronds divided 3–4 times, borne on a trunk; skirt of dead stipes and rachises; stipes with orange-brown scales; and sori away from lamina margin, with saucer-shaped indusia forming less than half a sphere. Plants to c. 2m tall can have only a weakly developed skirt, and can be difficult to separate from juvenile *Cyathea cunninghamii* – see the eFloraNZ.

Grows mostly in colder forests, especially at higher elevations.

Compare with *Cyathea colensoi, C. cunninghamii* and *Dicksonia fibrosa*.

Frond length: 1650–3000mm

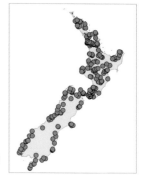

Endemic to Aotearoa
New Zealand.

Frond without much-shortened pinnae at base

Fronds

Stipes arising from creeping rhizome

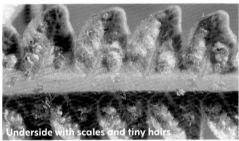

Underside with scales and tiny hairs

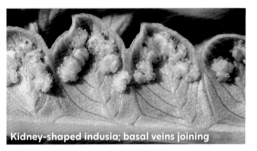

Kidney-shaped indusia; basal veins joining

Cyclosorus interruptus

THELYPTERIDACEAE

Distinguished by its scaly and obscurely hairy fronds, divided once but with deep secondary lobes, and spread along creeping rhizomes; pinnae not much shortened at base of frond; veins of adjacent secondary lobes joining; and sori with kidney-shaped indusia.

Grows in and around warmer wetlands, including in geothermal areas.

Compare with *Christella dentata* and *Thelypteris confluens*.

Frond length: 225–1450mm

Indigenous to Aotearoa New Zealand and many parts of the tropics and subtropics.

Large plant

Fronds divided once

Stipes tufted and scaly

Pinna underside with round indusia and netted veins

Young sori and indusia in irregular rows

Cyrtomium falcatum[+]

holly fern

DRYOPTERIDACEAE

Distinguished by its once-divided, tufted fronds with broad, somewhat sickle-shaped pinnae that have netted veins; and sori in many irregular rows, with round indusia.

Grows on the ground, often near the coast, and often near urban areas.

Compare with *Asplenium oblongifolium*.

Frond length: 200–1100mm

Naturalised in Aotearoa New Zealand. Indigenous to Asia.

Fronds

Frond divided three times

Sori with hood-like indusia partially covering sporangia

Frond segments with pointed apices

Stipes spread along creeping rhizome

Cystopteris fragilis[†]

CYSTOPTERIDACEAE

Distinguished by its broad fronds (up to 180mm wide) divided 2–3 times, usually spread along shortly creeping rhizomes; frond segments with somewhat pointed apices; and hood-like indusia.

Grows on the ground, usually near urban areas.

Frond length: 110–475mm

Naturalised in Aotearoa New Zealand. Indigenous to Europe, Asia and North America.

Frond divided twice; segments with blunt apices

Fronds

Sori with hood-like indusia, immature

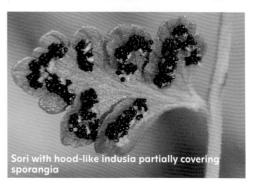

Sori with hood-like indusia partially covering sporangia

Cystopteris tasmanica

bladder fern

CYSTOPTERIDACEAE

Distinguished by its narrow fronds (usually < 75mm wide) divided 1–2 (rarely three) times, somewhat spread along shortly creeping rhizomes; frond segments with blunt apices; and hood-like indusia.

Grows in colder areas, under forest or in the open, often on base-rich rocks.

Frond length: 25–470mm

Indigenous to Aotearoa New Zealand and Australia.

Frond divided twice

Fronds

Sori elongated and diagonal to segment midvein

Grooves of axes not connected

Stipes arising from creeping rhizome

Deparia petersenii subsp. *congrua*

ATHYRIACEAE

Distinguished by its scaly and (inconspicuously) hairy fronds divided 2–3 times, spread along creeping rhizomes; grooves on the upperside of the rachis and costae not connected; and sori straight with indusia margins ragged.

Grows on damp ground in warmer places, often at forest margins.

Subspecies *petersenii* is a rare weed in Aotearoa – see the eFloraNZ.

Compare with *Diplazium australe*.

Frond length: 180–910mm

Probably indigenous to Aotearoa New Zealand, as well as Australia and many Pacific islands.

Fronds narrow, widest near their apex

Trunk thick, with skirt of dead fronds

Stipes short and greenish, hairy at base

Rachis hairy, bearing fertile pinnae

Sori at lamina margin

Dicksonia fibrosa

whekī-ponga

DICKSONIACEAE

Distinguished by its large, narrow fronds usually divided three times, borne on a trunk; skirt of dead fronds; short greenish or orange-brown stipes; hairy axes; and sori at the lamina margin. The trunk can become very wide.

Grows mostly in colder forest.

Compare with *Cyathea smithii*.

Frond length: 1100–3000mm

Endemic to Aotearoa
New Zealand.

Frond

Short trunk

Hairs on frond underside not in tufts

Stipes long, mostly orange-brown or grey

Sori at lamina margin

Dicksonia lanata subsp. *hispida*

tūākura, stumpy tree fern

DICKSONIACEAE

Distinguished by its large fronds divided 3–4 times, on a short trunk (to *c.* 2m tall); long and mostly orange-brown or grey stipes; covering of fine hairs on the undersides of the frond axes; and sori at the lamina margin.

Grows mostly in kauri forest.

Frond length: 500–2000mm

Endemic to Aotearoa New Zealand.

Frond

Hairs on frond underside in orange tufts

Stipes long and orange-brown

Trunkless plants growing in colony on hillside

Sori at lamina margin

Dicksonia lanata subsp. *lanata*

tūākura, creeping tree fern

DICKSONIACEAE

Distinguished by its large, more-or-less tufted fronds divided 3–4 times, without a trunk but with underground stolons producing new tufts; long and mostly orange-brown stipes; tufts of hairs at the costa junctions on the frond undersides; and sori at the lamina margin.

Grows mostly in colder forests, often at middle or higher elevations. It is often found on hillsides, where it can form extensive colonies.

Frond length: 400–2000mm

Endemic to Aotearoa New Zealand.

Stipes blackish

Growing in grove; with many orange-brown dead fronds

Buds on trunk

Axis undersides with bristly hairs; sori at lamina margin

Stipes dark brown, with brown hairs

Dicksonia squarrosa

whekī

DICKSONIACEAE

Distinguished by its large fronds divided 3–4 times, borne from a trunk; black or dark-brown stipes with many long brown hairs; brown, bristly hairs on axis undersides; and sori at the lamina margin. The orange-brown dead fronds are also characteristic. This is New Zealand's only tree fern that regularly re-sprouts from buds on the trunk.

Grows in forests, particularly on wetter ground. Often found in groves.

Frond length: 1250–2400mm

Endemic to Aotearoa New Zealand.

Fronds with axes repeatedly forking

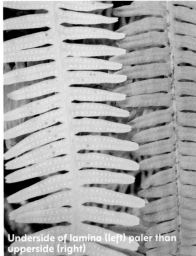

Underside of lamina (left) paler than upperside (right)

Axes below the forking branches mostly lacking lamina segments

Lamina underside, with sori and short red hairs

Dicranopteris linearis

GLEICHENIACEAE

Distinguished by its fronds with forking branches bearing lamina segments that are usually bluish-green below, 11–42mm long; mostly lacking lamina segments below the forks; scattered red hairs on the underside of the axes; and sori of 7–12 sporangia with many sori per lamina segment.

Grows among scrub in thermal areas from Te Moana-a-Toi Bay of Plenty to Lake Taupō.

Compare with *Sticherus* species.

Frond length: 140–1220mm

Indigenous to Aotearoa New Zealand and warmer parts of Africa, Asia, Australia and the Pacific Islands.

Frond divided three times

Grooves of axes connected, shallowly U-shaped

Sori elongated and diagonal to segment midvein

Fronds tufted; stipes scaly

Diplazium australe

ATHYRIACEAE

Distinguished by its tufted, scaly but not hairy fronds divided 3–4 times; grooves on the upperside of the rachis and costae connected and U-shaped; and sori usually straight with indusia margin entire or nearly so.

Grows on damp ground usually in warmer places, often at forest margins and in clearings.

Compare with *Athyrium filix-femina* and *Deparia petersenii.*

Frond length: 340–1400mm

Indigenous to Aotearoa
New Zealand and Australia.

Fronds divided twice

Stipes tufted, with golden-brown scales

Darkened rachis–costa junction; square-ended pinnae

Indusia kidney-shaped, domed when young

Dryopteris affinis[†]

golden male fern

DRYOPTERIDACEAE

Distinguished by its tufted fronds divided twice; golden-brown stipe scales; rather square-ended secondary pinnae; and kidney-shaped indusia that are domed when young. The rachis–costae junctions on the frond underside are prominently darkened when fresh. Can be difficult to distinguish from *Dryopteris filix-mas*.

Grows on the ground in the open and within exotic and native forest.

Frond length: 550–1290mm

Naturalised in Aotearoa New Zealand. Indigenous to Europe, Africa and Asia.

Frond divided three times

Fronds tufted

Elongated secondary pinnae pointing toward frond base

Sori (young) with kidney-shaped indusia

Stipe scales dark brown

Dryopteris dilatata[†]

broad buckler fern

DRYOPTERIDACEAE

Distinguished by its tufted fronds divided three times; dark-brown stipe scales; and kidney-shaped indusia. On the basal primary pinnae, the secondary pinnae pointing towards the frond base are elongated.

Grows on the ground in the open and within exotic and native forest.

Similar naturalised *Dryopteris* are rarely present – see the eFloraNZ.

Frond length: 330–960mm

Naturalised in Aotearoa New Zealand. Indigenous to Europe and Asia.

Frond divided twice

Indusia kidney-shaped, not domed when young

Stipe scales pale brown

Frond underside, with old sori

Dryopteris filix-mas†

male fern

DRYOPTERIDACEAE

Distinguished by its tufted fronds divided twice; pale-brown stipe scales; rather round-ended secondary pinnae; and kidney-shaped indusia that are not domed when young. The rachis–costae junctions on the frond underside are not prominently darkened when fresh. Can be difficult to distinguish from *Dryopteris affinis*.

Grows on the ground in the open and within exotic and native forest.

Frond length: 240–1190mm

Naturalised in Aotearoa New Zealand. Indigenous to Europe, Africa, Asia and North America.

Vegetative stem with whorled branches

Brown, unbranched fertile stems, each with apical cone

Multiple upright vegetative stems

Dominating what was mown grass

Equisetum arvense[+]

field horsetail

EQUISETACEAE

Distinguished by its upright, ribbed vegetative stems, usually < 500mm tall, with branches arising in whorls. Brown, unbranched fertile stems with a cone at their apex are produced in spring.

Grows usually in damp areas, such as beside watercourses. Spreads by underground rhizomes, and dense growth can exclude other vegetation. Occurrences outside Manawatū (where entrenched) should be reported to the local council.

Stem length: 100–800mm

Naturalised in Aotearoa New Zealand. Indigenous to Europe, Asia and North America.

Grove of tall, unbranched stems

Cone at apex of stem

Whorls of black-tipped leaves

Equisetum hyemale[†]

rough horsetail

EQUISETACEAE

Distinguished by its tall, upright, ribbed stems, to 2m high, usually with no branches but with whorls of small, black-tipped leaves. Cones are sometimes produced at the stem apex.

Grows usually near gardens, having spread by its underground rhizomes. Often on damp ground.

Stem length: 1000–2000mm

Naturalised in Aotearoa New Zealand. Indigenous to Europe, Asia and North America.

Fronds with new growth

Pouched lamina segments, densely scaly underneath

Branches usually short

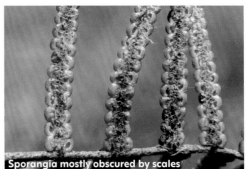

Sporangia mostly obscured by scales

Gleichenia alpina

alpine tangle fern

GLEICHENIACEAE

Distinguished by its pouched lamina segments < 1mm long, on side branches from forking branches; segment undersides obscured by scales; orange-brown (becoming pale), somewhat-triangular scales on axis undersides; and sori with two sporangia that sit on the lamina underside, lacking indusia, with one sorus per lamina segment. The fronds have shorter branches than the other *Gleichenia* species.

Grows in subalpine wetlands and scrub, and other cold, open habitats.

Frond length: 85–1250mm

Indigenous to Aotearoa New Zealand and Australia.

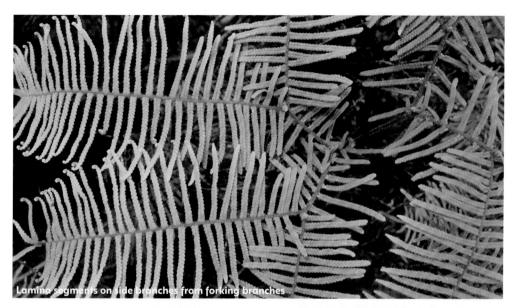

Lamina segments on side branches from forking branches

Pouched lamina segments with whitish undersides

Sori with two sporangia; mixture of scales

Gleichenia dicarpa

tangle fern

GLEICHENIACEAE

Distinguished by its usually pouched lamina segments < 2mm long, on side branches from forking branches; segments with whitish undersides; mixture of somewhat-triangular and star-shaped scales on axis undersides; and sori with usually two sporangia that sit on the lamina underside, lacking indusia, with one sorus per lamina segment.

Grows usually in low-fertility wetlands, or scrub on wet ground.

Hybridises frequently with *Gleichenia microphylla*, producing intermediates.

Frond length: 80–1460mm

Indigenous to Aotearoa New Zealand, Australia, New Caledonia and possibly south-eastern Asia.

Lamina segments on side branches from forking branches

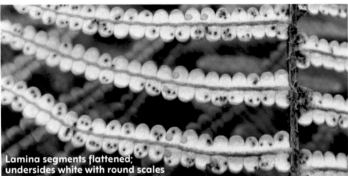

Lamina segments flattened; undersides white with round scales

Sori embedded in pits

Gleichenia inclusisora

pitted tangle fern

GLEICHENIACEAE

Distinguished by its flattened lamina segments < 2mm long, on side branches from forking branches; segments with white undersides that often bear prominent round scales; round scales sometimes on axis undersides; and sori with usually three sporangia that are embedded in pits on the lamina underside and lack indusia, with one sorus per lamina segment.

Grows in low-fertility wetlands, open scrub or short forest.

Frond length: 130–1230mm

Endemic to Aotearoa New Zealand.

Tangle of fronds

Lamina segments flat, green; sori mostly with 3–4 sporangia

Scales on axes mostly star-shaped

Gleichenia microphylla

waewae kākā, carrier tangle fern

GLEICHENIACEAE

Indigenous to Aotearoa New Zealand and Australia.

Distinguished by its flattened lamina segments < 3mm long, on side branches from forking branches; segments with green undersides and often pointed apices; mostly star-shaped, dark scales on axis undersides; and sori with usually 3–4 sporangia that sit on the lamina underside, lacking indusia, with one sorus per lamina segment.

Grows in scrub and open forest, particularly along natural or human-induced margins.

Hybridises frequently with *Gleichenia dicarpa*, producing intermediates.

Frond length: 110–1600mm

Frond

New fronds emerging from long-creeping rhizome

Pinnae sessile

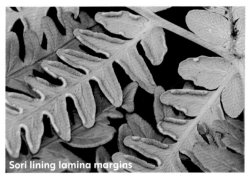

Sori lining lamina margins

Histiopteris incisa

mātātā, water fern

DENNSTAEDTIACEAE

Distinguished by its pale-green, glabrous fronds divided 2–4 times, spread along creeping rhizomes; sessile pinnae; netted veins, and sori lining the lamina margin.

Grows in open, often disturbed areas, including tracks through forest. Often found on damp ground.

Frond length: 170–2200mm

Indigenous to Aotearoa New Zealand, Australia, many Pacific islands, Asia, Africa and South America.

Frond with pinnae arising at c. 90°

Fronds spread along creeping rhizome

Indusia developing from margin

Lamina veins reaching margins at indentations

Hiya distans

Hypolepis distans

DENNSTAEDTIACEAE

Distinguished by its narrow fronds divided 2–3 times, spread along creeping rhizomes; lower pinnae connected to rachis at c. 90°; lamina veins reaching margins in small indentation; and almost circular sori on the lamina margin, with inrolled indusia. The polished red-brown stipe is distinctive.

Grows on wet ground, such as wetlands, often in the open but also under light forest; often on humus mounds or decomposing wood.

Compare with *Hypolepis* species.

Frond length: 120–1700mm

Indigenous to Aotearoa New Zealand and Australia.

Habit in open

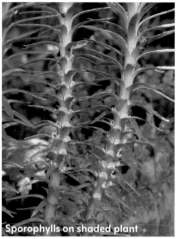

Sporophylls on shaded plant

Bulbils (brown) and sporophylls

Huperzia australiana

Lycopodium australianum
fir clubmoss

LYCOPODIACEAE

Distinguished by its sporophylls not being grouped into distinct cones; and tufted, erect stems. It is the only lycophyte in Aotearoa to have bulbils for vegetative propagation.

Grows on the ground in colder, open habitats; usually in uplands.

Stem length: 35–480mm

Indigenous to Aotearoa New Zealand, Australia, New Guinea and Indonesia.

Fronds among bryophytes

Fronds forked

Lamina margins with dark border and spines

Sori at apex of lamina segments

Hymenophyllum armstrongii

HYMENOPHYLLACEAE

Distinguished by its small fronds (≤ 32mm long), undivided or forked 1–3 times, and spread along creeping rhizomes; lamina margins with dark spines and often dark border; and solitary, sessile sorus with entire indusium at apex of lamina segment. Easily overlooked due to its small size, but distinctive.

Grows with mosses and liverworts, as an epiphyte on tree branches or matting damp rocks, even extending into the alpine zone.

Compare with *Hymenophyllum minimum.*

Frond length: 4–32mm

Endemic to Aotearoa
New Zealand.

Fertile frond

Lamina segments glabrous and with entire margins

Growing with bryophytes on rock above a creek

Hymenophyllum australe

Hymenophyllum atrovirens

HYMENOPHYLLACEAE

Distinguished by its often narrow fronds divided 2–4 times, spread along creeping rhizomes; glabrous lamina segments with entire margins; the flat wing that extends for most of the stipe's length; and its riparian habitat.

Grows in and around flowing water, usually on rock. Can be partially submerged.

Compare with *Hymenophyllum flexuosum* and *H. pluviatile*.

Frond length: 38–210mm

Indigenous to Aotearoa New Zealand and Australia.

Fertile frond, with sori not upturned

Lamina segments toothed

Hymenophyllum bivalve

HYMENOPHYLLACEAE

Distinguished by its fronds divided 4–5 times, often > 120mm long, and spread along creeping rhizomes; glabrous lamina segments with toothed margins; and sori lying in the same plane as the frond (rather than upturned).

Grows in forest, on the ground or as an epiphyte on tree trunks.

Compare with *Hymenophyllum multifidum*.

Frond length: 90–350mm

Indigenous to Aotearoa New Zealand and Australia.

Fronds growing from creeping rhizomes

Segments on both sides of primary pinnae

Lamina segments toothed; rachis winged

Indusia shallowly toothed

Hymenophyllum cupressiforme

HYMENOPHYLLACEAE

Distinguished by its fronds divided 2–3 times, ≤ 75mm long, and spread along creeping rhizomes; glabrous lamina segments with toothed margins; shallowly toothed indusial flaps; rachis winged for its entire length; and with secondary lamina segments on the frond's apical and basal sides of the primary pinnae.

Grows on the ground, most often in lighter and drier forest and scrub, below 760m elevation.

Compare with *Hymenophyllum peltatum* and *H. revolutum*.

Frond length: 15–75mm

Indigenous to Aotearoa New Zealand and Australia.

Fertile frond

Carpeting forest floor, via creeping rhizome

Lamina margins entire;
sori often paired and indusia triangular

Hymenophyllum demissum

irirangi

HYMENOPHYLLACEAE

Distinguished by its fronds divided 3–5 times, spread along creeping rhizomes; mostly glabrous lamina segments with entire margins; stipe not winged; and sori often in pairs at lamina segment apices, with often triangular indusia.

Grows on the ground and as an epiphyte in forest and tall scrub. Common, and the most likely *Hymenophyllum* species to be encountered in lowland forest, especially on the ground.

Frond length: 75–450mm

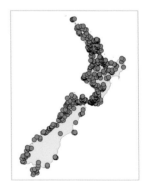

Endemic to Aotearoa
New Zealand.

Fertile frond with mature sori

Fertile frond; lamina segments broad

Lamina margins entire; sori round

Hymenophyllum dilatatum

matua mauku

HYMENOPHYLLACEAE

Distinguished by its fronds divided 3–4 times, spread along creeping rhizomes; mostly glabrous lamina segments with entire margins; partially winged stipe; and round sori that are partially immersed in the lamina segment apices. The broad lamina segments, 1.3–2.5mm wide, are distinctive.

Grows in forest and tall scrub, usually as an epiphyte but also on the ground.

Frond length: 60–570mm

Endemic to Aotearoa New Zealand.

Lamina glabrous, margins entire

Small frond with fan-shaped primary pinnae

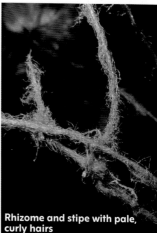

Rhizome and stipe with pale, curly hairs

Hymenophyllum flabellatum

fan-like filmy fern

HYMENOPHYLLACEAE

Distinguished by its fronds divided 3–4 times with somewhat fan-shaped primary pinnae, and spread along creeping rhizomes; mostly glabrous lamina segments with entire margins; pale, curly hairs on the rhizome, stipe and rachis; and stipe not winged.

Grows in forests, usually as an epiphyte, but also on banks and rock faces.

Compare with *Hymenophyllum rufescens*.

Frond length: 25–425mm

Indigenous to Aotearoa New Zealand, Australia, Vanuatu, Fiji, Sāmoa and French Polynesia.

Frond

Stipe with broad, flexuous wing

Lamina segments flexuous, with entire margins

Hymenophyllum flexuosum

HYMENOPHYLLACEAE

Distinguished by its fronds divided 3–5 times, spread along creeping rhizomes; lamina flexuous (wavy) with glabrous segments and entire margins; and a broad wing along its stipe and rachis that is usually prominently flexuous.

Grows in forest, on the ground or rocks, or as a low epiphyte.

Compare with *Hymenophyllum australe, H. pluviatile* and *H. pulcherrimum.*

Frond length: 70–400mm

Endemic to Aotearoa
New Zealand.

Frond

Many epiphytic fronds, connected by creeping rhizomes

Lamina with entire margins; sori immersed in segment apices

Star-shaped hairs on lamina

Hymenophyllum frankliniae

rusty filmy fern

HYMENOPHYLLACEAE

Distinguished by its fronds divided three (rarely four) times, spread along creeping rhizomes; and flattened lamina segments with entire margins and covered with star-shaped, pale red- or orange-brown hairs on both surfaces.

Grows in forest, usually as an epiphyte on tree fern trunks.

Compare with *Hymenophyllum malingii*.

Previously known as *Hymenophyllum ferrugineum*, but that species is confined to South America.

Frond length: 40–295mm

Endemic to Aotearoa New Zealand.

Frond fan-shaped, with immersed sori

Smaller frond

Lamina margin with forked hairs

Epiphytic on trunk of tree fern

Hymenophyllum lyallii

HYMENOPHYLLACEAE

Distinguished by its small, fan-shaped fronds divided 2–4 times, spread along creeping rhizomes; lamina segment margins with small teeth that bear forked hairs; and sori immersed in lamina segment apices. The sori distinguish this species from some similarly sized fan-shaped liverworts with which it is often confused.

Grows in forest, usually as an epiphyte.

Frond length: 9–90mm

Indigenous to Aotearoa New Zealand and Australia.

Fronds, with grey upperside

Star-shaped hairs, grey on upperside

Epiphytic on *Libocedrus bidwillii* trunk

Underside with reddish-brown hairs

Hymenophyllum malingii

HYMENOPHYLLACEAE

Distinguished by its fronds divided 3–4 times, spread along creeping rhizomes; almost cylindrical lamina segments with entire margins; and dense, star-shaped hairs that are grey on the lamina upperside and reddish-brown on the underside.

Grows usually in colder forest as an epiphyte on the trunks of *Libocedrus bidwillii* or *Metrosideros umbellata*.

Compare with *Hymenophyllum frankliniae*.

Frond length: 40–360mm

Endemic to Aotearoa
New Zealand.

Forming a mat in a rock crevice, under rhizome of *Lecanopteris pustulata*

Fronds with solitary sorus at rachis apex

Indusia with spines on outer surface

Hymenophyllum minimum

HYMENOPHYLLACEAE

Distinguished by its small fronds (≤ 30mm long) divided 1–2 times, spread along creeping rhizomes; glabrous lamina segments with toothed margins; solitary, stalked sorus at the rachis apex; and indusia that have spines on their outer surfaces.

Grows often on rocks, but can also grow on banks and as an epiphyte, ranging from the coast through forest and scrub to the alpine zone.

Compare with *Hymenophyllum armstrongii*.

Endemic to Aotearoa New Zealand.

Frond length: 6–30mm

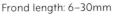

Frond

Fronds epiphytic from creeping rhizome on branch

Lamina segments with toothed margins; sori upturned

Blanketing upland forest

Hymenophyllum multifidum

much-divided filmy fern

HYMENOPHYLLACEAE

Distinguished by its fronds usually divided 4–5 times, often > 120mm long, and spread along creeping rhizomes; mostly glabrous lamina segments with toothed margins; and sori upturned at nearly 90° from the rest of the frond.

Grows widely, from shaded forest as an epiphyte or on the ground, to exposed rocks, including in alpine areas, where it may be stunted.

Compare with *Hymenophyllum bivalve*.

Frond length: 13–380mm

Indigenous to Aotearoa New Zealand, Lord Howe Island and some Pacific islands.

Fertile frond

Fronds spread along creeping rhizome

Carpeting ground

Sori in marginal, cup-shaped indusia

Hymenophyllum nephrophyllum

Trichomanes reniforme, Cardiomanes reniforme
raurenga, kidney fern

HYMENOPHYLLACEAE

Distinguished by its kidney-shaped, almost round fronds, spread along creeping rhizomes; and sori in marginal, cup-shaped indusia.

Grows in forest and scrub, as an epiphyte or on the ground.

Frond length: 40–280mm

Endemic to Aotearoa
New Zealand.

Segments only on apical side of primary pinnae

Rachis winged; lamina margins toothed

Indusia entire

Hymenophyllum peltatum

HYMENOPHYLLACEAE

Distinguished by its fronds divided twice (rarely three times), ≤ 140mm long, and spread along creeping rhizomes; glabrous lamina segments with toothed margins; entire indusial flaps; rachis usually winged for its entire length; and secondary lamina segments only on the frond's apical side of the primary pinnae.

Grows on the ground or as an epiphyte in forest or alpine scrub. Tends to be in colder habitats, and often above 750m elevation.

Compare with *Hymenophyllum cupressiforme* and *H. revolutum*.

Frond length: 20–140mm

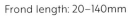

Indigenous to Aotearoa New Zealand, Australia, New Guinea, Borneo, South Africa and Chile.

Frond

Lamina segments with entire margins, slightly flexuous

Stipe with narrow wing

Hymenophyllum pluviatile

HYMENOPHYLLACEAE

Distinguished by its fronds divided 4–5 times, spread along creeping rhizomes; lamina slightly flexuous (wavy) with glabrous segments and entire margins; and a narrow stipe wing that extends almost to the base.

Grows on the ground or as a low epiphyte in forests with high rainfall.

Compare with *Hymenophyllum australe* and *H. flexuosum.*

Frond length: 75–190mm

Endemic to Aotearoa New Zealand.

Epiphytic on tree trunk

Fertile frond

Lamina segments with entire margins, flexuous

Stipes tufted and winged

Hymenophyllum pulcherrimum

tufted filmy fern

HYMENOPHYLLACEAE

Distinguished by its fronds divided 4–5 times, tufted from an erect or short-creeping rhizome; glabrous lamina segments with entire margins; prominently winged stipe; and flexuous (wavy) lamina. The only *Hymenophyllum* in Aotearoa with tufted fronds.

Grows in forest, usually as an epiphyte in colder habitats.

Compare with *Hymenophyllum flexuosum.*

Frond length: 120–720mm

Endemic to Aotearoa
New Zealand.

Frond

Fronds whitish when dry

Sori immersed in segment apices

Hymenophyllum rarum

HYMENOPHYLLACEAE

Distinguished by its fronds usually divided 2–3 times, spread along creeping rhizomes; glabrous lamina segments with entire margins; stipe not winged; and sori immersed within the lamina segment apices. The whitish-green colour of dry fronds is distinctive.

Grows in lowland forest to alpine scrub, usually as an epiphyte but rarely on the ground.

Frond length: 13–235mm

Indigenous to Aotearoa New Zealand and Australia.

Rachis lacking a wing except near its apex

Secondary segments on both sides of primary pinnae

Indusia deeply toothed

Hymenophyllum revolutum

HYMENOPHYLLACEAE

Distinguished by its fronds divided 2–3 times, ≤ 100mm long, and spread along creeping rhizomes; mostly glabrous lamina segments with toothed margins; deeply toothed indusial flaps; rachis winged only near its apex; and with secondary lamina segments on the apical and basal sides of the primary pinnae.

Grows in damp forest, on the ground or as an epiphyte. Mostly below 750m elevation.

Compare with *Hymenophyllum cupressiforme* and *H. peltatum*.

Endemic to Aotearoa New Zealand.

Frond length: 11–100mm

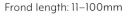

Stipe hairy

Sori partially immersed in lamina segments

Frond somewhat triangular; lamina upperside hairy

Hymenophyllum rufescens

HYMENOPHYLLACEAE

Distinguished by its somewhat triangular fronds divided 1–3 times, spread along creeping, hairy rhizomes; densely hairy lamina segments with entire margins; and stipes that are hairy but not winged.

Grows in forests as an epiphyte, or on banks and rocks, usually at colder sites.

Compare with *Hymenophyllum flabellatum*.

Frond length: 25–130mm

Endemic to Aotearoa New Zealand.

Indusia with crests

Frond with black, zigzagging rachis in the apical half

Stipe with narrow wing

Hymenophyllum sanguinolentum

piripiri

HYMENOPHYLLACEAE

Distinguished by its fronds usually divided 4 times, spread along creeping rhizomes; lamina segments with entire margins; narrowly winged stipe; rachis and costae undersides with usually only scattered hairs; and round sori with raised crests on the outer surfaces of the indusia.

Grows in forest and scrub, usually as an epiphyte but also on the ground. Generally found in warmer sites.

Compare with *Hymenophyllum villosum*; both have blackish, often zigzagging rachises and smell of blood when dry.

Frond length: 50–440mm

Indigenous to Aotearoa New Zealand and the Cook Islands.

Frond

Lamina olive-green

Stipe with bristly hairs

Hymenophyllum scabrum

rough filmy fern

HYMENOPHYLLACEAE

Distinguished by its olive-green fronds usually divided four times, spread along creeping rhizomes; lamina segments with scattered hairs and entire margins; and stipe not winged but with dense, bristly hairs (which can fall off in old fronds).

Grows in forest as an epiphyte or on the ground.

Frond length: 100–780mm

Endemic to Aotearoa New Zealand.

Frond with black, zigzagging rachis

Desiccated fronds, will unfurl with water

Stunted fronds in alpine habitat

Indusia lacking crests

Stipe hairy

Frond underside with hairy axes

Hymenophyllum villosum

hairy filmy fern

HYMENOPHYLLACEAE

Distinguished by its fronds usually divided four times, spread along creeping rhizomes; lamina segments with entire margins; narrowly winged hairy stipe; rachis and costae undersides moderately to densely hairy; and egg-shaped sori with indusia that lack crests.

Grows in forest and scrub, as an epiphyte or on the ground. Generally in colder sites, extending to alpine rock outcrops.

Compare with *Hymenophyllum sanguinolentum*; both have blackish, often zigzagging rachises and smell of blood when dry.

Endemic to Aotearoa New Zealand.

Frond length: 25–275mm

Frond

Mature sori; non-glandular hairs

Immature sori with weakly developed indusia

Hypolepis ambigua

common pig fern

DENNSTAEDTIACEAE

Distinguished by its fronds divided 2–4 times, bearing only non-glandular hairs, and spread along creeping rhizomes; broad lamina segments (> 1mm wide); and almost circular sori on the lamina margin, with only weakly developed, inrolled indusia.

Grows in open, often disturbed areas, including tracks through forest.

Compare with *Pteridium esculentum*, which is also common but has a thicker frond texture and oblong segments.

Frond length: 210–1850mm

Endemic to Aotearoa New Zealand.

Frond

Thick stipes arising from creeping rhizome

Many hairs glandular; inrolled indusia well developed over sori

Lamina upperside

Hypolepis dicksonioides

giant pig fern

DENNSTAEDTIACEAE

Distinguished by its large fronds divided 3–5 times, bearing glandular and non-glandular hairs, and spread along creeping rhizomes; stipe diameter often > 6mm; and almost circular sori on the lamina margin, with well-developed, inrolled indusia.

Grows in warmer areas, including thermal sites, usually in the open.

Frond length: 220–2450mm

Indigenous to Aotearoa New Zealand, Norfolk Island and several Pacific islands.

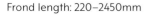

Lamina upperside usually whitish

Underside covered with glandular hairs; sori without inrolled indusia

Fronds broad

Hypolepis lactea

milky pig fern

DENNSTAEDTIACEAE

Distinguished by its usually broad, whitish fronds divided 2–4 times, spread along creeping rhizomes, and bearing colourless glandular and non-glandular hairs that are usually sticky to the touch; and almost circular sori near the lamina margin, usually with no indusia. The lamina upperside sometimes exudes a milky substance.

Grows in and around forest, usually in disturbed areas including along tracks.

Endemic to Aotearoa New Zealand.

Compare with *Hypolepis rufobarbata*. *Hypolepis amaurorhachis* is also similar but occurs only in the far south – see the eFloraNZ.

Frond length: 240–1900mm

Frond

Open subalpine habitat

Lamina divided into narrow segments; indusia slightly inrolled

Hypolepis millefolium

thousand-leaved fern

DENNSTAEDTIACEAE

Distinguished by its fronds divided 3–5 times, bearing only non-glandular hairs, and spread along creeping rhizomes; narrow lamina segments (< 1mm wide); and almost circular sori on the lamina margin, with only slightly inrolled indusia. The finely divided fronds are distinctive.

Grows in colder areas, extending to subalpine sites, often in the open but also along tracks through forest.

Frond length: 120–1000mm

Endemic to Aotearoa
New Zealand.

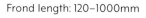

Fronds often narrow

Lamina margins fringed by reddish hairs; sori without inrolled indusia

Hypolepis rufobarbata

sticky pig fern

DENNSTAEDTIACEAE

Distinguished by its often narrow fronds divided 2–4 times, spread along creeping rhizomes, and bearing glandular as well as abundant red-brown (colourless when young) non-glandular hairs on the lamina surfaces and margins; and almost circular sori near the lamina margin, usually with no indusia. The red hairs fringing the lamina margin are distinctive.

Grows within cooler, wetter forest, often in disturbed areas including along tracks.

Compare with *Hypolepis lactea.*

Frond length: 115–1180mm

Endemic to Aotearoa New Zealand.

Several plants rooted in mud underwater

Leaves tufted from corm-like stem

Sporangia within swollen leaf bases

Isoetes alpina & *Isoetes kirkii*

alpine quillwort & quillwort

ISOETACEAE

Distinguished by their undivided leaves that have conspicuous air chambers, and are tufted from a corm-like stem; and the sporangia formed in the swollen leaf bases. Two species have been recognised but their distinctiveness is uncertain. Plants from Te Ika-a-Māui North Island and Te Waipounamu South Island may correspond to *Isoetes kirkii* and *I. alpina* respectively – see the eFloraNZ.

Grow submerged in lakes and tarns, rooted in mud, sand or stones.

Leaf length: 25–450mm

Both endemic to Aotearoa New Zealand.

Frond

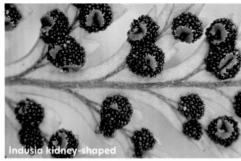

Indusia kidney-shaped

Stipe scales bristle-like

Frond upperside; axes with bristle-like scales

Stipes spread along shortly creeping rhizome

Lastreopsis hispida

bristly shield fern

DRYOPTERIDACEAE

Distinguished by its fronds divided 3–4 (rarely 5) times, somewhat spread along shortly creeping rhizomes; frond axes densely covered in dark, bristle-like scales as well as smaller hairs; and kidney-shaped indusia.

Grows on the ground in forest and scrub, usually in damper sites.

Compare with *Leptolepia novae-zelandiae*.

Frond length: 265–1400mm

Indigenous to Aotearoa
New Zealand and Australia.

Fronds

Frond upperside hairy

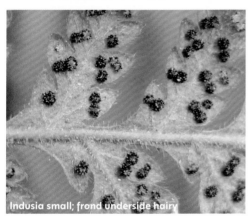

Indusia small; frond underside hairy

Frond underside with old sori

Lastreopsis velutina

velvet fern

DRYOPTERIDACEAE

Distinguished by its tufted fronds divided 3–4 times, with a dense covering of soft hairs on both surfaces; and small, kidney-shaped indusia. The frond is velvety to the touch.

Grows on the ground in forest and scrub, usually in drier sites such as hillsides.

Compare with *Parapolystichum* species.

Frond length: 240–960mm

Endemic to Aotearoa New Zealand.

Frond deeply once-lobed

Sori round or oval, without indusia

Rhizome with spreading orange-brown scales

Lecanopteris novae-zealandiae

Microsorum novae-zealandiae, Zealandia novae-zealandiae

POLYPODIACEAE

Distinguished by its bright-green, once-lobed fronds with netted veins, spread along thick, creeping rhizomes (diameter usually > 4mm) that have spreading, orange-brown scales; and large round or oval sori without indusia.

Grows as an epiphyte, usually within upland forests.

Frond length: 150–900mm

Endemic to Aotearoa New Zealand.

Fertile fronds (at right) with narrower lobes

Carpeting the ground

Thick, creeping rhizome with appressed, broad scales

Sori round or oval, without indusia

Lecanopteris pustulata

Microsorum pustulatum, Zealandia pustulata
kōwaowao, hound's tongue fern

POLYPODIACEAE

Distinguished by its bright-green, once-lobed (rarely undivided) fronds with netted veins, spread along thick, creeping rhizomes (diameter usually > 4mm) that have appressed, blackish, broad scales; and large round or oval sori without indusia. Fertile fronds have narrower lobes than sterile fronds.

Grows on the ground or as an epiphyte (in the open or within forest) from the coast to the uplands, often in drier habitats.

Compare with *Polypodium vulgare*.

Frond length: 45–750mm

Indigenous to Aotearoa
New Zealand and Australia.

Fronds dull green

Carpeting the ground (mostly juvenile fronds)

Thin, creeping rhizome with spreading, narrow scales

Sori small, without indusia

Lecanopteris scandens

Microsorum scandens, Dendroconche scandens
mokimoki, fragrant fern

POLYPODIACEAE

Distinguished by its dull-green, once-lobed fronds (undivided when juvenile) with netted veins, spread along wiry, creeping rhizomes (usually < 4mm diameter) that have spreading, blackish, narrow scales; and small round or oval sori without indusia.

Grows as an epiphyte or on the ground, usually within warmer, wetter forest.

Frond length: 45–620mm

Indigenous to Aotearoa New Zealand and Australia.

Frond

Stipe arising from long-creeping rhizome

Indusia opening towards lamina margin

Frond upperside

Leptolepia novae-zelandiae

lace fern

DENNSTAEDTIACEAE

Distinguished by its fronds divided 3–5 times, spread along creeping rhizomes; and somewhat circular sori near the lamina margin, protected by indusia that open towards the margin. The secondary pinnae closest to the rachis facing the frond base are shorter than those facing the frond apex.

Grows within shaded forest, usually in deep soils.

Compare with *Lastreopsis hispida*.

Frond length: 135–1150mm

Endemic to Aotearoa New Zealand.

Frond with long stipe, and pinnae at base of frond not much shorter than those above

Lamina segments flat

Sporangia not in sori of regular shape and size

Leptopteris hymenophylloides

heruheru, single crepe fern

OSMUNDACEAE

Distinguished by its translucent, membranous, dull-green, somewhat-triangular fronds divided 3–4 times, which are tufted, sometimes forming a short trunk; lamina segments flattened in the plane of the frond; basal pinnae not much shorter than those above; sporangia not in discrete sori of regular shape and size; and long stipes (≥ 110mm long). The erect rhizome distinguishes even small plants from most Hymenophyllaceae filmy ferns.

Grows on the ground in forest.

Frond length: 220–980mm

Endemic to Aotearoa New Zealand.

Lamina segments bending upwards

Shortened pinnae at frond base; stipes short

Sporangia not in sori of regular shape and size

Leptopteris superba

ngutu kākāriki, Prince of Wales feathers fern

OSMUNDACEAE

Distinguished by its translucent, membranous, dark-green, narrowly elliptic fronds divided 3–4 times, which are tufted, sometimes forming a short trunk; lamina segments bent at c. 90° to the plane of the frond; basal pinnae much shortened; sporangia not in discrete sori of regular shape and size; and short stipes (≤ 190mm long). The erect rhizome distinguishes even small plants from most Hymenophyllaceae filmy ferns.

Grows on the ground in usually colder, wet forest.

Frond length: 250–1000mm

Endemic to Aotearoa New Zealand.

Prostrate sterile fronds

Prostrate sterile fronds and erect fertile fronds

Indusia jagged and opening outwards

Lindsaea linearis

LINDSAEACEAE

Distinguished by its dimorphic, once-divided fronds, at least somewhat spread along creeping rhizomes; prostrate sterile fronds and erect fertile fronds; fan-shaped lamina segments; and jagged indusia that open outwards.

Grows usually on poor soils in scrub, but also under open forest.

Compare with *Asplenium flabellifolium* and *A. trichomanes.*

Frond length: 50–700mm

Indigenous to Aotearoa New Zealand, Australia and New Caledonia.

Frond divided three times

Indusia curved along rounded lamina segments

Fronds growing from creeping rhizomes

Lindsaea trichomanoides

LINDSAEACEAE

Distinguished by its non-dimorphic fronds divided 2–3 times, at least somewhat spread along creeping rhizomes; lamina segments with rounded apices; and entire indusia that curve around the lamina margin and open outwards.

Grows in forest or scrub, on the ground and at the base of trees.

Frond length: 70–440mm

Indigenous to Aotearoa
New Zealand and Australia.

Frond with many tertiary segments

Fronds hanging from bank above stream

Indusia rectangular, not curving; opening outward

Lindsaea viridis

LINDSAEACEAE

Distinguished by its non-dimorphic fronds divided 2–4 times, tufted or spread along creeping rhizomes; lamina segments with blunt apices; and mostly entire indusia that are rectangular on the lamina margin and open outwards.

Grows usually beside streams and waterfalls, but also on damp ground in forest.

Frond length: 40–400mm

Endemic to Aotearoa New Zealand.

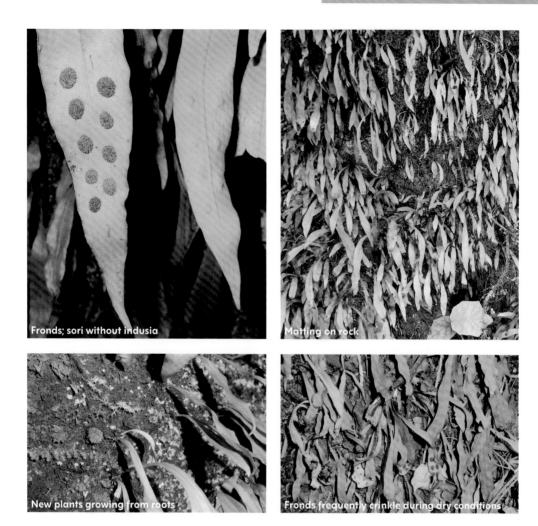

Fronds; sori without indusia

Matting on rock

New plants growing from roots

Fronds frequently crinkle during dry conditions

Loxogramme dictyopteris

Anarthropteris lanceolata
wharengārara, lance fern

POLYPODIACEAE

Distinguished by its undivided, glabrous, tufted fronds; and large round or oval sori without indusia. The proliferous roots produce new plants, and may be mistaken for a creeping rhizome.

Grows as a low epiphyte or on rocks, usually in warmer, lowland forest.

Compare with *Lecanopteris scandens* and *Notogrammitis billardierei.*

Frond length: 40–360mm

Endemic to Aotearoa
New Zealand.

Frond divided four times

Fronds

Underside white; sori in tube-like indusia

Underside green; immature sori

Loxsoma cunninghamii

LOXSOMATACEAE

Distinguished by its glabrous fronds divided 3–4 times, with pale-green or white undersides, and spread along creeping rhizomes; and sori on the lamina margins in tube-like indusia.

Grows in forest, often at the margins or on banks and streamsides.

Frond length: 370–1270mm

Endemic to Aotearoa
New Zealand.

Horizontal stems looping above ground

Upright stem with downward-curling branch apices

Cones sessile and pendulous from branch apices

Lycopodiella cernua

Lycopodium cernuum, Palhinhaea cernua

LYCOPODIACEAE

Distinguished by its sessile, pendulous cones; erect, much-branched stems with downward-curling branch-endings; and horizontal stems looping above ground.

Grows in open areas, often on banks. Especially common around thermal areas and in the north.

Stem height: 50–620mm

Indigenous to Aotearoa New Zealand and many parts of the tropics and subtropics.

Prostrate habit, with mature cones

Sessile cones lateral on prostrate stems

Prostrate *Lycopodiella diffusa* (left), with erect *L. lateralis* (right)

Lycopodiella diffusa

Lycopodium diffusum, L. ramulosum,
  Lateristachys diffusa
carpet clubmoss

LYCOPODIACEAE

Distinguished by its lateral (rather than terminal), mostly sessile cones; and usually prostrate growth habit.

Grows in wet, often open habitats. It generally occurs in colder places than the related *Lycopodiella lateralis*.

Stem length: 25–170mm

Indigenous to Aotearoa
New Zealand and Australia.

Cones lateral on stems rather than at stem apex

Semi-shaded stems with mature cones

Upright habit

Lycopodiella lateralis

Lycopodium laterale, Lateristachys lateralis

LYCOPODIACEAE

Distinguished by its lateral (rather than terminal), mostly sessile cones; and usually erect, sparingly branched growth habit.

Grows in damp, often open habitats. It generally occurs in warmer places than the related *Lycopodiella diffusa*.

Stem height: 50–700mm

Indigenous to Aotearoa New Zealand, Australia and New Caledonia.

Cones erect and stalked, from prostrate, horizontal stems

Maturing cone at centre

Lycopodiella serpentina

Lycopodium serpentinum, Pseudolycopodiella serpentina, Brownseya serpentina
bog clubmoss

LYCOPODIACEAE

Distinguished by its erect cones on unbranched stalks that arise from prostrate horizontal stems.

Grows in wetlands within limited areas of Te Tai Tokerau Northland and Waikato.

Stem height: 15–140mm

Indigenous to Aotearoa New Zealand, Australia and New Caledonia.

Cones erect from ends of upward-pointing branches

Growing as a colony

Juvenile with spreading leaves

Lycopodium deuterodensum

Pseudolycopodium densum
puakarimu

LYCOPODIACEAE

Distinguished by its erect, sessile cones at the apex of upward-pointing branches from erect, much-branched stems; underground horizontal stems; and appressed leaves of adults. Juveniles have spreading leaves.

Grows in colonies in forest and scrub, often on poor soils.

Stem height: 150–1900mm

Indigenous to Aotearoa New Zealand, Australia and New Caledonia.

Cones erect and stalked; leaves arranged spirally

Tall plant (with *Lycopodium volubile* at back)

Small orange, prostrate plant

Among rocks in alpine zone

Lycopodium fastigiatum

Austrolycopodium fastigiatum
alpine clubmoss

LYCOPODIACEAE

Distinguished by its erect cones on stalks that can be branched; and spirally arranged, spreading leaves. Plants in exposed habitats can be short and orange; those under shelter are taller and greener.

Grows usually in colder habitats, including the alpine zone, but also at the margins of forest and scrub.

Stem length: 20–520mm

Indigenous to Aotearoa
New Zealand and Australia.

Scrambling plant with erect, stalked cones

Stem underside with central rows of smaller leaves

Stem upperside with flattened lateral leaves

Lycopodium scariosum

Diphasium scariosum
creeping clubmoss

LYCOPODIACEAE

Distinguished by its erect, stalked, solitary or paired cones; and scrambling stems that have their larger leaves flattened into one plane, with rows of smaller leaves on the underside.

Grows in open habitats, usually near forest or scrub, and in colder areas.

Stem length: 45–600mm

Indigenous to Aotearoa New Zealand, Australia, New Guinea, Borneo and Philippines.

Sterile branches with flattened leaves

Scrambling plant with pendulous cones on fertile branches

Stem underside with central row of smaller leaves

Stem upperside with central rows of smaller leaves

Lycopodium volubile

Pseudodiphasium volubile
waewaekoukou, climbing clubmoss

LYCOPODIACEAE

Distinguished by its pendulous cones on often extensive fertile branches; and scrambling and climbing stems that have their larger leaves flattened into one plane, and smaller leaves on both the upper- and undersides.

Grows usually at the margins of forest and scrub, or under a light canopy.

Stem length: often extensive

Indigenous to Aotearoa New Zealand, New Caledonia, Solomon Islands, Vanuatu , New Guinea and south-eastern Asia.

Climbing axes that repeatedly fork

Tangle of wire-like axes

Sterile lamina segments paler underneath

Fertile lamina segments bearing sporangia on marginal lobes

Lygodium articulatum

mangemange, bushman's mattress

LYGODIACEAE

Distinguished by its climbing, twining habit, with wire-like axes that repeatedly fork; dimorphic sterile and fertile lamina segments; and sporangia borne on marginal lobes of fertile lamina segments.

Grows in forest and scrub, often on margins and in canopy gaps.

Frond length: often extensive

Endemic to Aotearoa New Zealand.

Grove of plants; fronds are tufted but proliferous roots mean that plants often grow in dense groves

Spherical tubers on stolons from rhizome

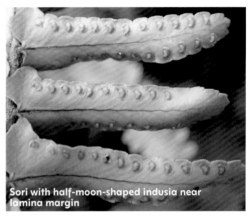

Sori with half-moon-shaped indusia near lamina margin

Nephrolepis cordifolia[+]

tuber ladder fern

NEPHROLEPIDACEAE

Distinguished by its narrow (30–100mm wide), once-divided fronds that are tufted from erect rhizomes, which produce long-creeping, proliferous stolons and spherical tubers; and sori with indusia shaped like a half-moon.

Grows on the ground, forming dense patches around settlements and at forest margins.

Frond length: 260–1130mm

Naturalised in Aotearoa New Zealand. Indigenous to many tropical areas, including north-eastern Australia and some Pacific islands.

Grove of plants

Sori with half-moon-shaped indusia near lamina margin

Fronds divided once

Nephrolepis flexuosa

NEPHROLEPIDACEAE

Distinguished by its narrow (18–55mm wide), once-divided fronds that are tufted from erect rhizomes, which produce long-creeping, proliferous stolons but no tubers; and sori with indusia shaped like a half-moon.

Grows on the ground around thermal areas from Te Moana-a-Toi Bay of Plenty to Lake Taupō.

Frond length: 200–750mm

Indigenous to Aotearoa New Zealand and probably some Pacific islands.

Sori absent from large area of frond apex

Epiphytic on trunk of tree; fronds undivided and tufted

Sori without indusia; lamina underside without hairs

Notogrammitis angustifolia

Grammitis magellanica

POLYPODIACEAE

Distinguished by its undivided, more-or-less tufted fronds; ill-defined, glabrous stipes; and oval sori without indusia or hairs. The sori are almost parallel to the lamina midvein, and often absent from a large portion of the frond apex.

Grows usually as an epiphyte above 1.5m from the ground, in wetter, colder forest.

Frond length: 15–220mm

Indigenous to Aotearoa New Zealand, Australia, South America and Africa.

Plants on bank

Stipes hairy, and more-or-less tufted

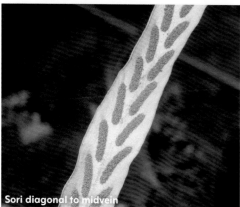

Sori diagonal to midvein

Sori without hairs

Notogrammitis billardierei

Grammitis billardierei
common strap fern

POLYPODIACEAE

Distinguished by its undivided, more-or-less tufted fronds; ill-defined, hairy stipes; and elongated sori without indusia or hairs. The long sori being diagonal to the midvein is characteristic.

Grows on the ground or as a low epiphyte (usually below 1.5m from the ground) in forest.

Similar to *Notogrammitis rigida*, which generally has bigger (up to 20mm wide), mostly glabrous fronds, and is confined to far southern coasts – see the eFloraNZ.

Frond length: 20–245mm

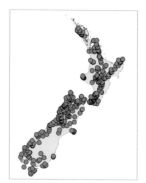

Indigenous to Aotearoa New Zealand and Australia.

Many plants growing on a bank, each with tufted fronds

Sori without indusia but with hairs

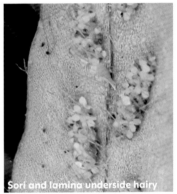

Sori and lamina underside hairy

Notogrammitis ciliata

Grammitis ciliata

POLYPODIACEAE

Distinguished by its undivided, more-or-less tufted fronds; ill-defined, hairy stipes; and oval or elongated sori without indusia but with pale hairs that can also be on the lamina underside. Rarely has glabrous fronds but still recognisable by its terrestrial, lowland habit and sori nearly parallel to the frond midvein.

Grows on the ground, usually on clay banks, but also on rocks in forest and scrub.

Frond length: 9–100mm

Endemic to Aotearoa New Zealand.

Fronds glabrous and with few sori clustered near apex

Young and old sori

Habitat: rock outcrop in alpine zone

Fronds

Rhizome creeping; stipes glabrous

Notogrammitis crassior

dwarf strap fern

POLYPODIACEAE

Distinguished by its small, undivided fronds (≤ 42mm long) spread along creeping rhizomes; ill-defined, glabrous stipes; and sori without indusia or hairs. The few sori are usually coalesced towards the frond apex.

Grows usually on rock in open alpine habitats.

Previously known, incorrectly, as *Grammitis poeppigiana*.

Frond length: 4–42mm

Indigenous to Aotearoa New Zealand, Australia, South America and some Southern Ocean islands.

Fronds

Sori, without indusia and hairs

Rhizome creeping; stipes hairy

Notogrammitis givenii

Grammitis givenii

POLYPODIACEAE

Distinguished by its undivided fronds usually spread along creeping rhizomes; clearly defined, hairy stipes; and sori without indusia or hairs.

Grows usually on rock, in open alpine habitats or in upland forest.

Frond length: 9–115mm

Endemic to Aotearoa New Zealand.

Young plants with barely lobed fronds

Fronds tufted from rhizome

Fronds

Oval sori without indusia

Notogrammitis heterophylla

Ctenopteris heterophylla
comb fern

POLYPODIACEAE

Distinguished by its tufted fronds divided 1–2 times; and oval sori without indusia or hairs.

Grows as an epiphyte or on banks and rocks, in the open or in forest.

Frond length: 22–440mm

Indigenous to Aotearoa
New Zealand and Australia.

Fronds

Rhizome long-creeping

Sori and lamina underside with dense red hairs

Sori and lamina underside with sparse red hairs

Notogrammitis patagonica

Grammitis patagonica

POLYPODIACEAE

Distinguished by its undivided fronds spread along creeping rhizomes; distinct stipes; and shortly elongated sori without indusia but with dark red-brown hairs that are also usually on the lamina underside.

Grows on the ground, often on rock or in the open, and usually at higher elevations or otherwise colder habitats.

Similar to the rare *Notogrammitis gunnii*, which has shorter fronds and paler hairs – see the eFloraNZ.

Frond length: 6–160mm

Indigenous to Aotearoa New Zealand and South America.

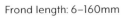

Epiphytic plants with fronds tufted from erect rhizomes

Frond margins undulate

Lamina underside hairy; sori elongated without indusia

Notogrammitis pseudociliata

Grammitis pseudociliata

POLYPODIACEAE

Distinguished by its undivided, more-or-less tufted fronds; ill-defined stipes; and elongated sori without indusia but with pale hairs that are also on the lamina underside. The frond margins are usually undulate and scalloped.

Grows as an epiphyte in usually wetter forest.

Frond length: 22–165mm

Indigenous to Aotearoa
New Zealand and Australia.

Sori

Young sori fringed with reddish hairs

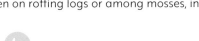

Several plants growing tufted in moss

Notogrammitis rawlingsii

Grammitis rawlingsii

POLYPODIACEAE

Distinguished by its undivided, more-or-less tufted fronds; ill-defined stipes; and elongated sori without indusia but fringed by dark red-brown hairs. These eyelash-like hairs are best seen when the sori are young.

Grows on the ground, often on rotting logs or among mosses, in northern forests.

Frond length: 31–165mm

Endemic to Aotearoa New Zealand.

Fronds with prostrate, undivided sterile parts and stalked, upright fertile parts

Fronds greener from growing under a canopy

Ophioglossum coriaceum

adder's tongue

OPHIOGLOSSACEAE

Distinguished by the lower leafy part of its frond being prostrate and undivided, to 14mm wide (rarely 20mm), with netted veins; and the unbranched, upright fertile part stalked with up to 17 (rarely 20) pairs of embedded sporangia at the apex.

Grows in open habitats such as grasslands, dunes and forest margins.

Similar to the rare *Ophioglossum petiolatum*, which has broader fronds and more sporangia – see the eFloraNZ.

Frond length: 12–250mm

Possibly endemic to Aotearoa New Zealand.

Sterile frond, divided twice

Fertile frond with modified apical pinnae

Young sporangia

Grove of large plants

Fronds tufted

Osmunda regalis[†]

royal fern

OSMUNDACEAE

Distinguished by its tufted, usually twice-divided fronds; and fertile fronds with dimorphic pinnae, with the fertile pinnae near the frond apex modified into clusters of sporangia.

Grows on wet ground, such as around wetlands and drains, usually in the open.

Frond length: 500–2070mm

Naturalised in Aotearoa New Zealand. Indigenous to Africa, Europe, Asia and the Americas.

Fertile frond

Sterile frond

Apical fertile frond segments somewhat triangular

Rachis zigzagging; sori (immature) lining lamina margins

Paesia scaberula

mātātā, ring fern

DENNSTAEDTIACEAE

Distinguished by its light-green fronds divided 3–4 times, with glandular hairs and zigzagging rachis, and spread along creeping rhizomes; and sori lining the lamina margin. The apical fertile frond segments are somewhat triangular.

Grows in open, often disturbed areas. Can cover large areas, such as reverting pasture.

Frond length: 150–1175mm

Endemic to Aotearoa New Zealand.

Frond divided once but with deep secondary lobes

Sometimes with a short trunk

Sori with no indusia; veins netted

Shorter pinnae at base of fronds

Dark veins of lamina

Pakau pennigera

Pneumatopteris pennigera
pākau, gully fern

THELYPTERIDACEAE

Distinguished by its more-or-less glabrous fronds, divided once but with deep secondary lobes, tufted from an erect rhizome, sometimes forming a short trunk; much-shortened pinnae at base of frond; veins of adjacent secondary lobes joining; and sori with no indusia. The dark veins are distinctive.

Grows usually in damp, shaded sites within forest, but not confined to gullies.

Frond length: 340–2070mm

Indigenous to Aotearoa New Zealand and Australia.

Frond

Stipes tufted from erect rhizome

Indusia kidney-shaped

Frond upperside with red hairs on axes

Rachis underside mostly glabrous

Parapolystichum glabellum

Lastreopsis glabella
smooth shield fern

DRYOPTERIDACEAE

Distinguished by its tufted fronds divided 3–4 times; rachis underside mostly lacking hairs; and kidney-shaped indusia. The hairs on the upperside of the axes are usually reddish.

Grows on the ground in forest and scrub, usually in damper sites.

Compare with *Lastreopsis* species.

Frond length: 120–900mm

Endemic to Aotearoa
New Zealand.

Frond

Rachis underside with many short hairs

Frond upperside with pale hairs

Stipes spread along creeping rhizome

Indusia kidney-shaped

Parapolystichum microsorum

Lastreopsis microsora
creeping shield fern

DRYOPTERIDACEAE

Distinguished by its fronds divided 3–4 times spread along creeping rhizomes; rachis underside densely covered in short, erect hairs; and kidney-shaped indusia. The hairs on the upperside of the axes are usually pale.

Grows on the ground in forest and scrub, usually in drier sites.

Compare with *Lastreopsis* species.

Frond length: 240–960mm

Indigenous to Aotearoa
New Zealand and Australia.

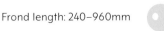

Fronds divided once; enlarged apical pinna on sterile fronds

Sori on lamina margins, joining at pinna apex

Rachis scales mostly appressed

Pellaea calidirupium

hot rock fern

PTERIDACEAE

Distinguished by its narrow fronds divided once, spread along creeping rhizomes; sori lining the lamina margins and usually joining at the pinna apex; and mostly appressed scales on the rachises and stipes. Sterile fronds often have an enlarged apical pinna.

Grows in dry, open, rocky areas.

Frond length: 35–460mm

Indigenous to Aotearoa New Zealand and Australia.

Fronds divided once

Sori not joining at pinna apex; pinnae pointed

Fronds spread along creeping rhizome

Many rachis scales spreading; pinnae rounded

Pellaea rotundifolia

tarawera, button fern

PTERIDACEAE

Distinguished by its narrow fronds divided once, spread along creeping rhizomes; sori lining the lamina margins that usually do not join at the pinna apex; and spreading scales on the rachises and stipes. Plants from New Zealand with long, pointed pinnae previously attributed to *Pellaea falcata* are included here.

Grows on the ground in forest, more so in the lowlands.

Compare with *Blechnum fluviatile*.

Frond length: 110–870mm

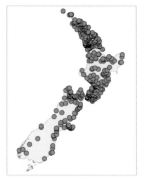

Indigenous to Aotearoa New Zealand and Norfolk Island.

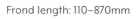

Pendulous epiphyte

Sporophylls and sterile leaves appressed

Phlegmariurus billardierei

LYCOPODIACEAE

Distinguished by its pendulous habit and pendulous cones that have appressed sporophylls. The cones connect to sterile stems with appressed leaves, which in turn lead into sterile stems with spreading leaves.

Grows epiphytically or on the ground, particularly on banks, usually in forest. It is more common in warmer, lowland habitats than *Phlegmariurus varius* but they often co-occur.

Stem length: 120–1800mm

Endemic to Aotearoa New Zealand.

Robust pendulous epiphyte

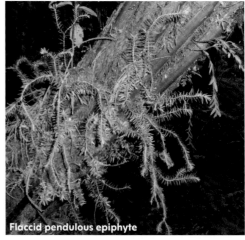

Flaccid pendulous epiphyte

Plant on ground with somewhat upright habit

Spreading sporophylls grading into spreading sterile leaves

Phlegmariurus varius

Lycopodium varium, Huperzia varia
whiri-o-Raukatauri, iwituna

LYCOPODIACEAE

Distinguished by its pendulous or erect habit; and cones that have at least some spreading sporophylls. The cones connect to sterile stems with spreading leaves, with no appressed sterile leaves near the cones. *Phlegmariurus varius* is more variable than *P. billardierei*.

Grows epiphytically or on the ground. Extends into colder habitats than *Phlegmariurus billardierei* but they often co-occur.

Stem length: 70–1000mm

Indigenous to Aotearoa
New Zealand and Australia.

Stalked cone surround by rosette of leaves

With developing cone

Phylloglossum drummondii

LYCOPODIACEAE

Distinguished by its stalked cone surrounded by a rosette of undivided leaves; and small size (≤ 50mm tall).

Grows on poor soils in gumland wetlands of Te Tai Tokerau Northland. Formerly extended to at least the Waikato. It appears above ground only in winter and spring, surviving summer as an underground tuber.

Stem length: 11–50mm

Indigenous to Aotearoa
New Zealand and Australia.

Leaves spread along creeping rhizome

Rooted in mud underwater

Young leaf with unfurling koru

Exposed at lake edge

Spherical sporangia casings

Pilularia novae-hollandiae

Pilularia novae-zealandiae
pillwort

MARSILEACEAE

Distinguished by its undivided, narrowly cylindrical leaves that are spread singly on a long-creeping rhizome; and sporangia borne in a hairy, spherical casing of 2–4mm diameter. Easily confused for a flowering plant except when sporangia casings or young leaves with unfurling koru are present.

Grows submerged in lakes and tarns, rooted in mud, sand or stones.

Frond length: 10–90mm

Indigenous to Aotearoa New Zealand and Australia.

Foliage fronds emerging from nest fronds

Foliage fronds pendulous from appressed nest fronds

Patches of sporangia on underside of lobe apices

Platycerium bifurcatum[†]

staghorn fern

POLYPODIACEAE

Distinguished by its strongly dimorphic fronds, with roundish, appressed fronds forming a 'nest' and pendulous foliage fronds that fork repeatedly. The foliage fronds are covered in star-shaped hairs and have large patches of sporangia at their apices when fertile.

Grows as an epiphyte, or rarely on walls. Commonly cultivated.

Frond length: 50–1050mm

Naturalised in Aotearoa New Zealand. Indigenous to Australia.

Frond divided once; most pinnae divided to rachis

Sori without indusia; pinna margins toothed

Creeping rhizome with orange-brown scales

Polypodium vulgare[+]

common polypody

POLYPODIACEAE

Distinguished by its once-divided fronds with netted veins, spread along thick, creeping rhizomes with orange-brown scales; pinna margins with small teeth; at least the basal frond segments divided to the rachis; and oval sori without indusia.

Grows on the ground, in the open or within scrub and forest.

Compare with *Lecanopteris pustulata*.

Frond length: 85–550mm

Naturalised in Aotearoa New Zealand. Indigenous to Europe, Africa and Asia.

Fronds divided three times

Habit

Stipe scales orange-brown

Indusia round, inflated; rachis scales orange-brown

Polystichum cystostegia

alpine shield fern

DRYOPTERIDACEAE

Distinguished by its tufted fronds divided 2–3 times; upperside of axes not darker than lamina; orange-brown rachis scales; secondary pinnae that are stalked only close to the rachis; and inflated, round indusia without a dark centre.

Grows on the ground, usually in open alpine areas above 900m elevation.

Frond length: 40–550mm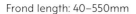

Endemic to Aotearoa New Zealand.

Frond

Upperside of axes darker than lamina

Rachis scales narrow; indusia round with dark centre

Polystichum neozelandicum

Polystichum richardii
pikopiko, black shield fern

DRYOPTERIDACEAE

Distinguished by its tufted fronds divided twice (barely three times); upperside of axes usually darker than lamina (not in eastern Otago); narrow, dark rachis scales; secondary pinnae that are stalked only close to the rachis; and flat, round indusia with a prominent dark centre.

Grows on the ground in forest and scrub.

Frond length: 140–1000mm

Endemic to Aotearoa
New Zealand.

Frond

Upperside of axes not darker than lamina

Rachis scales broad; indusia round with dark centre

Polystichum oculatum

DRYOPTERIDACEAE

Distinguished by its tufted fronds divided twice (barely three times); upperside of axes not darker than lamina; broad, dark or irregularly pale-brown rachis scales; secondary pinnae that are stalked only close to the rachis; and flat, round indusia with a prominent dark centre.

Grows on the ground in forest and scrub, often in dry sites.

Frond length: 125–780mm

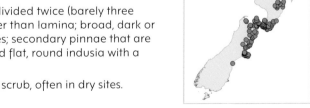

Endemic to Aotearoa New Zealand.

Frond

Rachis scales dark with pale margins; young sori with no indusia

Mature sori with no indusia

Polystichum sylvaticum

DRYOPTERIDACEAE

Distinguished by its tufted fronds divided 2–3 times; upperside of axes not darker than lamina; broad rachis scales with dark centres and pale margins; primary costae slightly winged; most secondary pinnae stalked; and indusia absent. Best identified when sori are young or just mature, as other species may lose their indusia on ageing fronds.

Grows on the ground in dark, damp forest.

Frond length: 150–800mm

Endemic to Aotearoa New Zealand.

Frond

Sprawling habit under canopy

Erect sward habit when in open

Indusia round with small dark centre

Rachis scales dark with pale margins

Secondary pinnae mostly stalked; upperside of axes not darker than lamina

Polystichum vestitum

pūniu, prickly shield fern

DRYOPTERIDACEAE

Distinguished by its rectangular, tufted fronds divided 2–3 times; upperside of axes not darker than lamina; broad rachis scales with dark centres and pale margins; primary costae not winged; most secondary pinnae stalked; and flat, round indusia with a small (or no) dark centre. Occasionally has a short trunk.

Grows on the ground from lowland forest to open alpine areas.

Frond length: 220–1700mm

Indigenous to Aotearoa New Zealand and Macquarie Island.

Frond

Upperside of axes darker than lamina

Rachis scales hairlike; indusia small (here crumpled)

Underside of frond

Polystichum wawranum

DRYOPTERIDACEAE

Distinguished by its tufted fronds divided twice (barely three times); upperside of axes darker than lamina; dark, hairlike rachis scales; secondary pinnae that are stalked only close to the rachis; and flat, round indusia with a small (or no) dark centre.

Grows on the ground in forest and scrub.

Frond length: 275–1270mm

Endemic to Aotearoa New Zealand.

Green stem repeatedly forking

Clusters of three fused sporangia

Psilotum nudum

fork fern

PSILOTACEAE

Distinguished by its green stems that repeatedly fork in two; inconspicuous, tiny leaves; and yellowish clusters of three fused sporangia.

Grows in warmer sites such as northern coasts or geothermal areas, usually on the ground, particularly around rocks, but also as a low epiphyte.

Stem length: 40–725mm

Indigenous to Aotearoa New Zealand and many tropical and subtropical regions.

Primary pinna of large frond

Axis winged towards its apex

Sori lining lamina margin

Pteridium esculentum

rārahu, bracken

DENNSTAEDTIACEAE

Distinguished by its dark-green, often large fronds divided 2–5 times, with inconspicuous non-glandular hairs, and spread along creeping rhizomes; and sori lining the lamina margin although often absent or difficult to see. The undivided lamina segments are distinctively narrowly oblong. The axes are partially winged.

Grows in open, often disturbed areas, including dry ground.

Compare with *Hypolepis* and *Pteris*.

Frond length: 250–2200mm

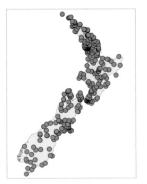

Indigenous to Aotearoa New Zealand, Australia, several Pacific islands, Central and South America and possibly Asia.

Frond

Lamina veins netted

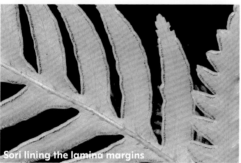

Sori lining the lamina margins

Basal secondary pinnae elongated

Pteris carsei

coastal brake

PTERIDACEAE

Distinguished by its large, tufted fronds divided 2–4 times; netted lamina veins; sori lining the lamina margins; secondary pinnae usually not stalked; and the basal secondary pinnae close to the rachis greatly elongated.

Grows in coastal forest.

Previously known as *Pteris comans*, but that species is confined to the south-west Pacific. Hybridises with *P. saxatilis*.

Frond length: 380–1430mm

Endemic to Aotearoa
New Zealand.

Fronds growing from creeping rhizomes

Sori lining the lamina margins

Fronds divided twice at base, elsewhere divided once

Veins free, forking but not netted

Pteris cretica[+]

Cretan brake

PTERIDACEAE

Distinguished by its fronds divided twice at their base and once elsewhere, spread along creeping rhizomes; free veins; and sori lining the lamina margins.

Grows in lowland habitats, usually near houses. Increasingly weedy in several parts of the country.

Frond length: 450–1320mm

Naturalised in Aotearoa New Zealand. Indigenous to Africa, Europe and Asia.

Frond, with many stalked secondary pinnae

Smaller frond

Sori lining the lamina margins

Lamina veins netted

Pteris macilenta

titipo, sweet fern

PTERIDACEAE

Distinguished by its tufted, often large fronds divided 3–5 times; netted lamina veins; sori lining the lamina margins; secondary pinnae usually stalked; basal secondary pinnae close to the rachis not greatly elongated; and pinnae not widely spaced.

Grows in forest and scrub, more so in warmer habitats.

Similar to hybrids between *Pteris carsei* and *P. saxatilis*.

Frond length: 200–1550mm

Endemic to Aotearoa New Zealand.

Frond with widely spaced pinnae

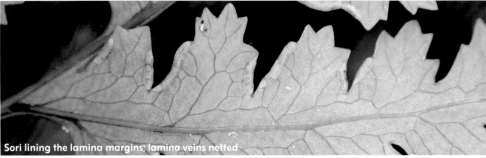

Sori lining the lamina margins; lamina veins netted

Pteris saxatilis

PTERIDACEAE

Distinguished by its tufted, often small fronds divided 2–4 times; netted lamina veins; sori lining the lamina margins; secondary pinnae usually stalked; basal secondary pinnae close to the rachis not greatly elongated; and pinnae widely spaced.

Grows in forest and scrub, usually near the coast.

Hybridises with *Pteris carsei*.

Frond length: 145–700mm

Endemic to Aotearoa
New Zealand.

Frond

Fronds tufted

Pinnae rectangular;
sori lining the lamina margins

Veins free, not netted

Pteris tremula

turawera, shaking brake

PTERIDACEAE

Distinguished by its tufted, often large fronds divided 3–4 times; free veins; and sori lining the margins of the lamina segments, which are somewhat rectangular.

Grows in lowland forest and scrub, and often invades gardens.

Similar to the indigenous *Pteris epaleata* and naturalised *P. dentata*, but both are very uncommon – see the eFloraNZ.

Frond length: 220–2440mm

Indigenous to Aotearoa New Zealand, Australia, Vanuatu and Fiji.

Frond divided once

Large plants on rock wall

Sori lining the lamina margins

Pteris vittata[+]

ladder brake

PTERIDACEAE

Distinguished by its fronds divided once, spread along creeping rhizomes; free veins; and sori lining the lamina margins.

Grows mostly in urban areas, on walls and rocks, but has also been found in a geothermal area.

Frond length: 95–1220mm

Naturalised in Aotearoa New Zealand. Indigenous to many other parts of the world.

Frond divided twice

Fronds tufted from bulbous rhizome

Primary costae swollen at junction with rachis

Sporangia in fused oblong clusters near margin

Ptisana salicina

Marattia salicina
para, king fern

MARATTIACEAE

Distinguished by its large, glossy fronds divided twice, tufted from a bulbous rhizome; swellings at the junctions of the primary costae with the rachis; and sporangia fused into oblong clusters just inside the lamina margin.

Grows in warmer forest, often in steep gullies out of reach of pigs.

Frond length: 2000–4000mm

Indigenous to Aotearoa New Zealand, Norfolk Island, Cook Islands and French Polynesia.

Fertile fronds longer, narrower; sterile fronds shorter, broader

Sori in several, ill-defined rows

Narrow, long-creeping rhizomes

Frond upperside with scattered star-shaped hairs

Frond underside densely covered with star-shaped hairs

Pyrrosia elaeagnifolia

ngārara wehi, leather-leaf fern

POLYPODIACEAE

Distinguished by its leathery or fleshy, undivided fronds spread along creeping rhizomes; lamina underside whitish and densely covered in star-shaped hairs, with similar hairs scattered on upperside; and sori without indusia and in multiple, ill-defined rows either side of the midvein.

Grows as an epiphyte or on rocks, usually where there are higher light levels. Often found on introduced trees.

Endemic to Aotearoa New Zealand.

Frond length: 20–185mm

Frond

Stipes spread along climbing rhizome; scales orange-brown

Indusia round

Rumohra adiantiformis

karuwhai, climbing shield fern

DRYOPTERIDACEAE

Distinguished by its leathery fronds divided 2–3 times spread along climbing or creeping rhizomes that are covered in orange-brown scales; and round indusia.

Grows usually as a climbing epiphyte in forest, but also scrambles over the ground.

Frond length: 85–1020mm

Indigenous to Aotearoa New Zealand, Australia, New Guinea, South America and Africa.

Habit

Combs of fertile pinnae short and narrow

Fronds with unbranched sterile axes

Schizaea australis

Microschizaea australis
southern comb fern

SCHIZAEACEAE

Distinguished by the sterile axes of its fronds being unbranched, smooth and narrow; and the fertile pinnae in 'combs' of 4–13 pairs, 3–17mm long, 1–4mm wide, without hairs among the sporangia.

Grows on damp, open ground, usually in colder places.

Similar to the generally bigger *Schizaea fistulosa* but can be difficult to distinguish in north-west Te Waipounamu South Island.

Frond length: 20–350mm

Indigenous to Aotearoa New Zealand, South America and possibly Tasmania.

Comb of fertile pinnae broad

Fronds with unbranched sterile axes

Hairs among the sporangia

Young fronds, some with forked sterile axes

Schizaea bifida

forked comb fern

SCHIZAEACEAE

Distinguished by the sterile axes of its fronds being rough, narrow, and unbranched or branched 1–3 times; and the fertile pinnae in 'combs' of 4–18 pairs, 5–24mm long, 3–10mm wide, with hairs among the sporangia.

Grows on poor soils, in scrub, open forest or wetlands.

Compare with the similarly sized *Schizaea fistulosa*, which usually has narrower combs.

Frond length: 50–650mm

Indigenous to Aotearoa New Zealand, Australia and New Caledonia.

Flattened axes forking several times

Young growth can be red

Fertile frond with combs terminating the flattened axes

Hairs among the sporangia

Schizaea dichotoma

fan fern

SCHIZAEACEAE

Distinguished by the sterile axes of its fronds being smooth, flattened and forking 3–7 times in a fan-like arrangement; and the fertile pinnae in 'combs' of 4–8 pairs, 2–8mm long, 1–6mm wide, with hairs among the sporangia.

Grows in forest and scrub, often on poor soils or thermally heated ground.

Frond length: 60–460mm

Indigenous to Aotearoa New Zealand and tropical and subtropical regions from Africa through Asia to the Pacific.

Fronds with unbranched sterile axes

Comb of fertile pinnae long and narrow

Young fronds

Schizaea fistulosa

Microschizaea fistulosa
comb fern

SCHIZAEACEAE

Distinguished by the sterile axes of its fronds being unbranched, smooth and narrow; and the fertile pinnae in 'combs' of 7–30 pairs, 9–38mm long, 2–5mm wide, without hairs among the sporangia.

Grows on poor soils, in scrub, open forest or wetlands.

Similar to the generally smaller *Schizaea australis* but can be difficult to distinguish in north-west Te Waipounamu South Island.

Frond length: 115–570mm

Indigenous to Aotearoa New Zealand, Australia, south-eastern Asia, New Guinea, and many Pacific islands.

Delicate leaves, with those of the central rows smaller

Green cone upright above stem

Matting the forest floor

Selaginella kraussiana[†]

African clubmoss

SELAGINELLACEAE

Distinguished by its largely prostrate, creeping stems; delicate leaves with a single, unbranched vein, and dimorphic with the leaves of the side rows larger than those in the centre; and often-inconspicuous green cones.

Grows on the ground, usually in damp, shady sites, including the sides of tracks and streams. Forms smothering mats, and spreads readily from vegetative fragments.

Stem length: 25–420mm

Naturalised in Aotearoa New Zealand. Indigenous to Africa.

Fronds with forking branches; branches often curving

Lamina segments paler underneath; prominent scales on undersides of axes; sori with 3–5 sporangia

Lamina segments present between forks

Sticherus cunninghamii

waekura, umbrella fern

GLEICHENIACEAE

Distinguished by its fronds with forking branches often curving, bearing lamina segments that are usually whitish or blue-green below, 5–20mm long; with lamina segments between the forks; prominent scales on the underside of the axes; and sori of 3–5 sporangia with many sori per lamina segment.

Grows in forest and scrub.

Frond length: 140–980mm

Endemic to Aotearoa New Zealand.

Frond with forking branches and lamina segments between forks

Underside green; scales hairlike

Fronds fan-shaped, with branches almost straight

Sticherus flabellatus

GLEICHENIACEAE

Distinguished by its fronds with forking branches bearing green lamina segments 14–55mm long; with lamina segments between the forks; pale, narrow or hairlike scales on the underside of the axes; and sori of 3–5 sporangia with many sori per lamina segment.

Grows in scrub, sparse forest or other open habitats.

Similar Te Waipounamu South Island plants will be *Sticherus tener* or *S. urceolatus* – see the eFloraNZ.

Frond length: 180–1490mm

Indigenous to Aotearoa New Zealand, Australia, New Guinea and New Caledonia.

Frond without much-shortened pinnae at base

Costae scaly; sori with kidney-shaped indusia

Veins free, not joining between secondary pinnae, which are divided nearly to primary costae

Thelypteris confluens

marsh fern

THELYPTERIDACEAE

Distinguished by its scaly but seemingly hairless fronds, divided twice, and spread along creeping rhizomes; pinnae not much shortened at base of frond; veins free; and sori with kidney-shaped indusia.

Grows in and around warmer wetlands, including in geothermal areas.

Compare with *Christella dentata* and *Cyclosorus interruptus*.

Frond length: 380–1150mm

Indigenous to Aotearoa New Zealand and tropical and subtropical regions from Africa through Asia to Australia.

Leaves mostly with tapering apices

Leaves dull;
pairs of sporangia with rounded ends

Tmesipteris elongata

PSILOTACEAE

Distinguished by its dull, narrow and usually tapering leaves that are spirally arranged or flattened towards the stem apex; and pairs of fused sporangia that are oblong with rounded ends. The only species in Aotearoa that can branch twice or more.

Grows in forest on the trunks of tree ferns and flowering plants, and from epiphytic *Astelia*.

Stem length: 70–1400mm

Indigenous to Aotearoa
New Zealand and Australia.

Leaves mostly with blunt or concave apices and hair-points

Leaves with blunt apices; sporangia with somewhat pointed ends

Pendulous stems

Tmesipteris horomaka

PSILOTACEAE

Distinguished by its spirally arranged, narrow leaves that have blunt or concave apices with hair-points; and pairs of fused sporangia with somewhat pointed ends that sit mostly flat on their leaf. Morphologically intermediate between *Tmesipteris elongata* and *T. tannensis*.

Grows in forest on the trunks of tree ferns.

Confined to the Port Hills and Horomaka Banks Peninsula, where it is the most common *Tmesipteris*.

Stem length: 60–510mm

Endemic to Aotearoa New Zealand.

Leaves broad, shiny, and flattened towards stem apex

Sporangia produced only near stem base

Tmesipteris lanceolata

PSILOTACEAE

Distinguished by its shiny, broad leaves that become flattened into one plane towards the stem apex; and pairs of fused sporangia that are spherical with rounded or slightly pointed ends. The sporangia are produced only in the basal part of the stem.

Grows in forest on the trunks of tree ferns.

Stem length: 30–185mm

Indigenous to Aotearoa New Zealand and New Caledonia.

Leaves spirally arranged, narrow with a slight S-shape

Pairs of rounded sporangia

Tmesipteris sigmatifolia

PSILOTACEAE

Distinguished by its spirally arranged, shiny, narrow and usually slightly S-shaped leaves; and pairs of fused sporangia that are spherical with rounded ends. The hair-points at the leaf apices are comparatively long.

Grows in forest on the trunks of tree ferns. Usually associated with kauri forest.

Stem length: 40–270mm

Indigenous to Aotearoa New Zealand and New Caledonia.

Pairs of sporangia with pointed ends

Epiphytic on trunk of ponga tree fern

Leaves usually with blunt apices and long hair-points

Tmesipteris tannensis

PSILOTACEAE

Distinguished by its spirally arranged, often narrow (occasionally broad), usually shiny leaves that usually have blunt apices with hair-points; and pairs of fused sporangia with pointed ends that rise away from their leaf.

Grows in forest and scrub on the trunks of tree ferns and seed-plant trees, or from epiphytic *Astelia*, and is the only Aotearoa species of *Tmesipteris* to regularly grow on the ground.

Stem length: 50–830mm

Endemic to Aotearoa New Zealand.

Frond divided twice

Fronds tufted from erect rhizome

Frond underside

Sporangia densely packed

Todea barbara

hard todea

OSMUNDACEAE

Distinguished by its tough, mostly glabrous, twice-divided fronds that are tufted, sometimes forming a short trunk; and sporangia densely covering the underside of pinnae towards the base of fertile fronds.

Grows on the ground, in the open or under light shade, often on poor soils.

Frond length: 210–2200mm

Indigenous to Aotearoa New Zealand, Australia and southern Africa.

Fronds growing from creeping rhizome on bank

Lamina segments and indusia both stalked

Rachis not winged; indusia tubular and stalked

Trichomanes colensoi

Polyphlebium colensoi

HYMENOPHYLLACEAE

Distinguished by its fronds divided three times, spread along creeping rhizomes; stipe and rachis not winged; mostly glabrous and stalked lamina segments with entire margins and veins forking 1–2 times; and tubular indusia that are often stalked.

Grows in dark forest on damp rock and banks, usually beside waterways.

Frond length: 35–150mm

Endemic to Aotearoa
New Zealand.

Frond

Frond underside, with long bristles from sori

Fronds tufted from rhizome

Lamina segments with veins forking several times

Trichomanes elongatum

Abrodictyum elongatum
bristle fern

HYMENOPHYLLACEAE

Distinguished by its tufted, broadly triangular, dark-green fronds divided 2–3 times; rachis mostly not winged; broad, mostly glabrous lamina segments with entire margins and the veins forking several times; and tubular indusia, often with a long projecting bristle.

Grows in forest on the ground, often in deep shade.

Frond length: 95–360mm

Endemic to Aotearoa
New Zealand.

Fronds matting on rock

Frond divided twice; rachis winged; lamina segments and tubular indusia both sessile

Trichomanes endlicherianum

Polyphlebium endlicherianum

HYMENOPHYLLACEAE

Distinguished by its fronds usually divided twice, spread along creeping rhizomes; winged stipe and rachis; sessile, mostly glabrous lamina segments with entire margins and a single, unbranched vein; and sessile, tubular indusia.

Grows in forest, usually on damp rock, but occasionally as a low epiphyte.

Frond length: 22–130mm

Indigenous to Aotearoa New Zealand, Australia, New Guinea, Borneo and many Pacific islands.

Fronds

Lamina segments with single, unbranched veins

Rachis winged

Fronds tufted from rhizome

Trichomanes strictum

Abrodictyum strictum
erect bristle fern

HYMENOPHYLLACEAE

Distinguished by its tufted, narrowly elliptic, medium-green fronds usually divided three times; winged rachis; narrow, mostly glabrous lamina segments with entire margins and a single, unbranched vein; and tubular indusia, often with a long projecting bristle.

Grows in forest on the ground, often in wetter, colder sites.

Frond length: 65–350mm

Endemic to Aotearoa
New Zealand.

Segments with many-branched veins

Tubular indusia, with long bristle (mature)

Epiphytic on trunk of tree fern

Trichomanes venosum

Polyphlebium venosum
veined filmy fern

HYMENOPHYLLACEAE

Distinguished by its light-green fronds divided 1–2 times, spread along creeping rhizomes; stipe not winged and rachis mostly not winged; mostly glabrous lamina segments with entire margins and many-branched veins; and tubular indusia.

Grows in forest, usually as an epiphyte, especially on tree fern trunks, but rarely on banks and rocks.

Frond length: 25–165mm

Indigenous to Aotearoa
New Zealand and Australia.

Gleichenia microphylla, page 151

Azolla pinnata, page 98.

Appendices
Acknowledgements
About the authors
Index

Appendices

APPENDIX 1: The genera within each family

Only families and genera included in this guide are listed here. The family for each genus is also given in the **Species profiles.**

FAMILY	GENERA	PLANT GROUP
Aspleniaceae	*Asplenium*	ferns
Athyriaceae	*Athyrium, Deparia, Diplazium*	ferns
Blechnaceae	*Blechnum*	ferns
Cyatheaceae	*Cyathea*	ferns
Cystopteridaceae	*Cystopteris*	ferns
Dennstaedtiaceae	*Histiopteris, Hiya, Hypolepis, Leptolepia, Paesia, Pteridium*	ferns
Dicksoniaceae	*Dicksonia*	ferns
Dryopteridaceae	*Cyrtomium, Dryopteris, Lastreopsis, Parapolystichum, Polystichum, Rumohra*	ferns
Equisetaceae	*Equisetum*	ferns
Gleicheniaceae	*Dicranopteris, Gleichenia, Sticherus*	ferns
Hymenophyllaceae	*Hymenophyllum, Trichomanes*	ferns
Isoetaceae	*Isoetes*	lycophytes
Lindsaeaceae	*Lindsaea*	ferns
Loxsomataceae	*Loxsoma*	ferns
Lycopodiaceae	*Huperzia, Lycopodiella, Lycopodium, Phlegmariurus, Phylloglossum*	lycophytes
Lygodiaceae	*Lygodium*	ferns
Marattiaceae	*Ptisana*	ferns
Marsileaceae	*Pilularia*	ferns
Nephrolepidaceae	*Nephrolepis*	ferns
Ophioglossaceae	*Botrychium, Ophioglossum*	ferns
Osmundaceae	*Leptopteris, Osmunda, Todea*	ferns
Polypodiaceae	*Lecanopteris, Loxogramme, Notogrammitis, Platycerium, Polypodium, Pyrrosia*	ferns
Psilotaceae	*Psilotum, Tmesipteris*	ferns
Pteridaceae	*Adiantum, Anogramma, Cheilanthes, Pellaea, Pteris*	ferns
Salviniaceae	*Azolla*	ferns
Schizaeaceae	*Schizaea*	ferns
Selaginellaceae	*Selaginella*	lycophytes
Tectariaceae	*Arthropteris*	ferns
Thelypteridaceae	*Christella, Cyclosorus, Pakau, Thelypteris*	ferns

APPENDIX 2: Species indigenous to Aotearoa New Zealand not included in this guide

See the eFloraNZ for more details.

SPECIES	DISTRIBUTION IN AOTEAROA
Arachniodes aristata	Rangitāhua Kermadec Islands only
Asplenium chathamense	Rēkohu Chatham Islands only
Asplenium pauperequitum	Tawhiti Rahi Poor Knights Islands and Rēkohu Chatham Islands only
Asplenium shuttleworthianum	Rangitāhua Kermadec Islands only
Blechnum kermadecense	Rangitāhua Kermadec Islands only
Blechnum neohollandicum	Very uncommon in Te Tai Tokerau Northland
Botrychium lunaria	Very uncommon in Te Waipounamu South Island mountains
Cyathea kermadecensis	Rangitāhua Kermadec Islands only
Cyathea milnei	Rangitāhua Kermadec Islands only
Davallia tasmanii subsp. *cristata*	One site in Te Tai Tokerau Northland
Davallia tasmanii subsp. *tasmanii*	Manawatāwhi Three Kings Islands only
Hymenophyllum polyanthos	Rangitāhua Kermadec Islands only
Hypolepis amaurorhachis	Subantarctic islands and very uncommon in southern Te Waipounamu South Island
Macrothelypteris torresiana	Rangitāhua Kermadec Islands and one (former) site in Te Tai Tokerau Northland
Nephrolepis brownii	Rangitāhua Kermadec Islands only
Notogrammitis gunnii	Uncommon in Te Waipounamu South Island mountains
Notogrammitis rigida	Subantarctic islands, Rakiura Stewart Island and Te Rua o Te Moko Fiordland
Ophioglossum petiolatum	Uncommon in Te Ika-a-Māui North Island and northern Te Waipounamu South Island
Parapolystichum kermadecense	Rangitāhua Kermadec Islands only
Pteris epaleata	Very uncommon in Te Rua o Te Moko Fiordland
Pyrrosia serpens	Rangitāhua Kermadec Islands only
Sticherus tener	Very uncommon in western Te Waipounamu South Island
Sticherus urceolatus	Very uncommon in western Te Waipounamu South Island
Trichomanes caudatum	Rangitāhua Kermadec Islands and very uncommon in northern Te Ika-a-Māui North Island
Trichomanes humile	Rangitāhua Kermadec Islands only

APPENDIX 3: Exotic species with 'naturalised' status in Aotearoa New Zealand not included in this guide

See the eFloraNZ for more details. The eFloraNZ also includes species with only a 'casual' naturalisation status.

SPECIES	DISTRIBUTION IN AOTEAROA NEW ZEALAND
Davallia griffithiana	Tāmaki Makaurau Auckland, Kirikiriroa Hamilton, Whanganui
Dryopteris cycadina	Te Tai Tokerau Northland, Waikato, Te Whanganui-a-Tara Wellington
Marsilea mutica	Tāmaki Makaurau Auckland, Paekākāriki, Kaiapoi
Polystichum proliferum	Whangārei, Tāmaki Makaurau Auckland, Waikato, Whanganui, Te Whanganui-a-Tara Wellington, Waitohi Picton, Ōtepoti Dunedin
Polystichum setiferum	Kirikiriroa Hamilton
Pteris dentata	Tāmaki Makaurau Auckland, Kirikiriroa Hamilton
Salvinia ×molesta	Te Tai Tokerau Northland, Tāmaki Makaurua Auckland, Waikato

APPENDIX 4: Collective names for genera and families, te reo Māori and English

Some colloquial names apply to multiple species, such as a family or genus, or part thereof. There are examples in te reo Māori and English, and some are given here. Only some groups have a colloquial name for a given language, and this list is not exhaustive.

GROUP OF SPECIES	TE REO	ENGLISH
Adiantum	huruhuru tapairu, makawe tapairu, tawatawa	maidenhair ferns
Asplenium	petako-pāraharaha, petako rauriki	spleenworts
Blechnum		hard ferns
Equisetum		horsetails
Gleichenia		tangle ferns
Hymenophyllaceae		filmy ferns
Hymenophyllum	mauku	
Hypolepis		pig ferns
Isoetes		quillworts
Lycopodiaceae	mātukutuku, tarakupenga, whareatua	clubmosses
Nephrolepis		ladder ferns
Notogrammitis		strap ferns
Polystichum		shield ferns
Pteris		brake ferns
Schizaea		comb ferns
Selaginella		spikemosses
Sticherus		umbrella ferns
Tmesipteris		fork ferns

Acknowledgements

I would like to acknowledge Wellington Botanical Society members and the fern enthusiasts on the iNaturalist website for shaping and testing the knowledge presented in this book; Bill Campbell, Jack Warden and Maureen Young for help locating some species; Jane Harris for making early drafts available for testing via Te Papa's website; and Te Papa Press for their guidance through publication: publisher Michael Upchurch, project editors Olivia Nikkel and Olive Owens, editor Teresa McIntyre, proofreader Matt Turner, designer Sarah Elworthy and imaging technician Jeremy Glyde. Thanks to Janet Hunt for supplying the diagrams and to Jean-Claude Stahl for the cover photographs.

This book was inspired by Cheng-Wei Chen's *Lycophytes and Ferns of the Solomon Islands*. The content is based on the Ferns and Lycophytes series of the online Flora of New Zealand, which was produced from a funding partnership coordinated by Te Papa's Patrick Brownsey and Manaaki Whenua Landcare Research's Ilse Breitwieser. The plant specimen collections of New Zealand's herbaria were the basis for this work, particularly those of Auckland Museum, Manaaki Whenua Landcare Research and Te Papa.

Many thanks to the generous support of the Deane Endowment Trust that helped fund the book.

Special thanks to Lara Shepherd for her long support. She has been dragged to see far more ferns than she could ever want to.

Leon Perrie, March 2024

About the authors

LEON PERRIE is Curator Botany at Te Papa and was the lead science curator for Te Papa's long-term exhibition *Te Taiao | Nature*. Leon specialises in plant taxonomy and the collection and curation of plant specimens. His research focuses on New Zealand's ferns: their numbers, locations and identification, and using DNA analyses to understand how ferns are related to one another and to species overseas. He was a contributing author for the Ferns and Lycophytes series for the online Flora of New Zealand.

PATRICK BROWNSEY was Curator Botany at the National Museum of New Zealand and Te Papa for over forty years, and expert with New Zealand ferns and lycophytes. He was the lead author for the Ferns and Lycophytes series for the online Flora of New Zealand. At the time of writing, he was Research Associate Botany and had previously been Head of Natural History at the museum where he also curated the philately (stamps) collection. Pat passed away in late 2023 and this book is a dedication to his work.

Index

First published in New Zealand in 2024 by Te Papa Press,
PO Box 467, Wellington, New Zealand
www.tepapapress.co.nz

Text: Leon Perrie and Patrick Brownsey © Museum of New Zealand Te Papa Tongarewa

All images by Leon Perrie apart from pages 4–7 and 30–31 by Jean-Claude Stahl
© Museum of New Zealand Te Papa Tongarewa.

TE PAPA® is the trademark of the Museum of New Zealand Te Papa Tongarewa

Te Papa Press is an imprint of the Museum of New Zealand Te Papa Tongarewa

A catalogue record is available from the National Library of New Zealand

ISBN 978-1-99-116555-8

Design by Sarah Elworthy
Diagrams by Janet Hunt
Digital imaging by Jeremy Glyde
Printed by Everbest Printing Investment Limited, China

Cover photographs by Jean-Claude Stahl © Museum of New Zealand Te Papa Tongarewa
Front cover: *Hymenophyllum demissum*
Back cover: *Cyathea medullaris* and other tree ferns

Te Papa Press acknowledges the generous support of the Deane Endowment Trust
to help fund this publication

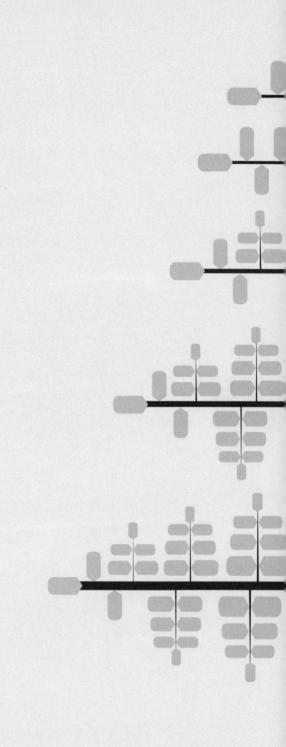